Mathematical Principles of Signal Processing

Springer
New York
Berlin
Heidelberg
Barcelona
Hong Kong
London
Milan
Paris
Singapore
Tokyo

Pierre Brémaud

Mathematical Principles of Signal Processing

Fourier and Wavelet Analysis

Springer

Pierre Brémaud
École Polytechnique Fédérale de Lausanne
Switzerland

and

INRIA/École Normale Supérieure
France
bremaud@ens.fr

Library of Congress Cataloging in Publication Data
Brémaud, Pierre.
Mathematical principles of signal processing / Pierre Brémaud.
p. cm.
Includes bibliographical references and index.

1. Signal processing—Mathematics. I. Title.
TK5102.9 .B72 2001
621.382′2′0151—dc21 2001042957

Printed on acid-free paper.

ISBN 978-1-4419-2956-3

Production managed by Allan Abrams; manufacturing supervised by Joe Quatela.

9 8 7 6 5 4 3 2 1

Springer-Verlag New York Berlin Heidelberg
A member of BertelsmannSpringer Science+Business Media GmbH

To Marion

Contents

Preface

Fourier theory is one of the most useful tools in many applied sciences, particularly, in physics, economics, and electrical engineering. Fourier analysis is a well-established discipline with a long history of successful applications, and the recent advent of wavelets is the proof that it is still very alive. This book is an introduction to Fourier and wavelet theory illustrated by applications in communications. It gives the mathematical principles of signal processing in such a way that physicists and electrical engineers can recognize the familiar concepts of their trade.

The material given in this textbook establishes on firm mathematical ground the field of signal analysis. It is usually scattered in books with different goals, levels, and styles, and one of the purposes of this textbook is to make these prerequisites available in a single volume and presented in a unified manner.

Because Fourier analysis covers a large part of analysis and finds applications in many different domains, the choice of topics is very important if one wants to devise a text that is both of reasonable size and of meaningful content. The coloration of this book is given by its potential domain of applications—signal processing. In particular, I have included topics that are usually absent from the table of contents of mathematics texts, for instance, the z-transform and the discrete Fourier transform among others.

The interplay between Fourier series and Fourier transforms is at the heart of signal processing, for instance in the sampling theory at large (including multiresolution analysis). In the classical Fourier theory, the formula at the intersection of the Fourier transform and the Fourier series is the Poisson formula. In mathematically oriented texts, it appears as a corollary or as an exercise and in most cases receives little attention, whereas in engineering texts, it appears under its avatar, the formula giving the Fourier transform of the Dirac comb. For obscure reasons, it is believed that the Poisson sum formula, which belongs to *classic* analysis, is too difficult, and students are gratified with a result of distributions theory that requires from them a higher degree of mathematical sophistication. Surprisingly, in the applied literature, whereas distribution theory is implicitly assumed to be innate, the basic properties of the Lebesgue integral, such as the dominated convergence and the Fubini theorem, are never stated precisely and seldom used, although these tools are easy to understand and would certainly answer many of the questions that alert students are bound to ask. In order to correct this unfortunate tradition, which has a demoralizing effect on good students, I have insisted on the fact that the classical Poisson formula is all that is needed in signal processing to justify the Dirac symbolism, and I have devoted some time and space to introduce the Lebesgue integral in a concise appendix, giving the precise statements of the indispensable tools.

The contents are organized in four chapters. Part A contains the Fourier theory in L^1 up to the classical results on pointwise convergence and the Poisson sum formula. Part B is devoted to the mathematical foundations of signal processing. Part C gives the Fourier theory in L^2. Finally, Part D is concerned with the time–frequency issue, including the Gabor transform, wavelets, and multiresolution analysis. The mathematical prerequisites consist of a working knowledge of the Lebesgue integral, and they are reviewed in the appendix.

Although the book is oriented toward the applications of Fourier analysis, the mathematical treatment is rigorous, and I have aimed at maintaining a balance between practical relevance and mathematical content.

Acknowledgments

Michael Cole translated and typed this book from a French manuscript, and Claudio Favi did the figures. Jean-Christophe Pesquet and Martin Vetterli encouraged me with stimulating discussions and provided the illustrations of wavelet analysis. They also checked and corrected parts of the manuscript, together with Guy Demoment and Emre Telatar. Sébastien Allam and Jean-François Giovanelli were always there when TEX tried to take advantage of my incompetence. To all of them, I wish to express my gratitude, as well as to Tom von Foerster, who showed infinite patience with my promises to deliver the manuscript on time.

Gif sur Yvette, France Pierre Brémaud
May 2, 2001

Part A

Fourier Analysis in L^1

Introduction

In 1807 Joseph Fourier (1768–1830) presented a solution of the heat equation[1]

$$\frac{\partial \theta}{\partial t} = \kappa \frac{\partial^2 \theta}{\partial^2 x},$$

where $\theta(x, t)$ is the temperature at time t and at location x of an infinite rod, and κ is the heat conductance. The initial temperature distribution at time 0 is given:

$$\theta(x, 0) = f(x).$$

(The solution of the heat equation is derived in Section A1·1.)

In fact, Fourier considered a circular rod of length, say, 2π, which amounts to imposing that the functions $x \to f(x)$ and $x \to \theta(x, t)$ are 2π-periodic. He gave the solution when the initial temperature distribution is a trigonometric series

$$f(t) = \sum_{n \in \mathbb{Z}} c_n e^{int}.$$

Fourier claimed that his solution was general because he was convinced that all 2π-periodic functions can be expressed as a trigonometric series with the coefficients

$$c_n = c_n(f) = \frac{1}{2\pi} \int_0^{2\pi} f(t) e^{-int} \, dt.$$

[1]The definitive form of his work was published in *Théorie Analytique de la Chaleur*, Firmin Didot éd., Paris, 1822.

Special cases of trigonometric developments were known, for instance, by Leonhard Euler (1707–1783), who gave the formula

$$\frac{1}{2}x = \sin(x) - \frac{1}{2}\sin(2x) + \frac{1}{3}\sin(3x) - \cdots,$$

true for $-\pi < x < +\pi$. But the mathematicians of that time were skeptical about Fourier's general conjecture. Nevertheless, when the propagation of heat in solids was set as the topic for the 1811 annual prize of the French Academy of Sciences, they surmounted their doubts and attributed the prize to Fourier's memoir, with the explicit mention, however, that it lacked rigor. Fourier's results that were in any case true for an initial temperature distribution that is a finite trigonometric sum, and be it only for this, Fourier fully deserved the prize, because his proof uses the general tricks (for instance, the differentiation rule and the convolution–multiplication rule) that constitute the powerful toolkit of Fourier analysis.

Nevertheless, the mathematical problem that Fourier raised was still pending, and it took a few years before Peter Gustav Dirichlet[2] could prove rigorously, in 1829, the validity of Fourier's development for a large class of periodic functions. Since then, perhaps the main guideline of research in analysis has been the consolidation of Fourier's ingenious intuition.

The classical era of Fourier series and Fourier transforms is the time when the mathematicians addressed the basic question, namely, what are the functions admitting a representation as a Fourier series? In 1873 Paul Dubois-Reymond exhibited a *continuous* periodic function whose Fourier series *diverges* at 0. For almost one century the threat of painful negative results had been looming above the theory. Of course, there were important positive results: Ulisse Dini[3] showed in 1880 that if the function is locally Lipschitz, for instance differentiable, the Fourier series represents the function. In 1881, Camille Jordan[4] proved that this is also true for functions of locally bounded variation. Finally, in 1904 Leopold Féjer[5] showed that one could reconstruct any continuous periodic function from its Fourier coefficients. These results are reassuring, and for the purpose of applications to signal processing, they are sufficient.

However, for a pure mathematician, the itch persisted. There were more and more examples of periodic continuous functions with a Fourier series that diverges at at least one point. On the other hand, Féjer had proven that if convergence is taken in the Césaro sense, the Fourier series of such continuous periodic function converges to the function at all points.

---

[2] Sur la convergence des séries trigonométriques qui servent à représenter une fonction arbitraire entre des limites données, *J. reine und angewan. Math.*, 4, 157–169.

[3] Serie di Fourier e altre rappresentazioni analitiche delle funzioni di une variabile reale, Pisa, Nistri, vi + 329 p.

[4] Sur la série de Fourier, *CRAS Paris*, 92, 228–230; See also *Cours d'Analyse de l'École Polytechnique*, I, 2nd ed., 1893, p. 99.

[5] Untersuchungen über Fouriersche Reihen, *Math. Ann.*, 51–69.

Outside continuity, the hope for a reasonable theory seemed to be completely destroyed by Nikolaï Kolmogorov,[6] who proved in 1926 the existence of a periodic locally Lebesgue-integrable function whose Fourier series *diverges at all points*! It was feared that even continuity could foster the worst pathologies. In 1966 Jean-Pierre Kahane and Yitzhak Katznelson[7] showed that given any set of null Lebesgue measure, there exists a continuous periodic function whose Fourier series diverges at all points of this preselected set.

The case of continuous functions was far from being elucidated when Lennart Carleson[8] published in the same year an unexpected result: Every periodic locally square-integrable function has an almost-everywhere convergent Fourier series. This is far more general than what the optimistic party expected, since the periodic continuous functions are, in particular, locally square-integrable. This, together with the Kahane–Katznelson result, completely settled the case of continuous periodic functions, and the situation finally turned out to be not as bad as the 1873 result of Dubois–Reymond seemed to forecast.

In this book, the reader will not have to make her or his way through a jungle of subtle and difficult results. Indeed, for the traveler with practical interests, there is a path through mathematics leading directly to applications. One of the most beautiful sights along this road may be Siméon Denis Poisson's[9] sum formula

$$\sum_{n\in\mathbb{Z}} f(n) = \sum_{n\in\mathbb{Z}} \hat{f}(n),$$

where f is an integrable function (satisfying some additional conditions to be made precise in the main text) and where

$$\hat{f}(\nu) = \int_{\mathbb{R}} f(t)\, e^{-2i\pi\nu t}\, dt$$

is its Fourier transform, where $\mathbb{R}$ is the set of real numbers. This striking formula found very nice applications in the theory of series and is one of the theoretical results founding signal analysis. The Poisson sum formula is the culminating result of Part A, which is devoted to the classical Fourier theory.

---

[6] Une série de Fourier–Lebesgue divergente partout, *CRAS Paris*, 183, 1327–1328.

[7] Sur les ensembles de divergence des séries trigonométriques, *Studia Mathematica*, 26, 305–306.

[8] Convergence and growth of partial sums of Fourier series, *Acta Math.*, 116, 135–157.

[9] Sur la distribution de la chaleur dans les corps solides, *J. École Polytechnique*, 19ème Cahier, XII, 1–144, 145–162.

A1

Fourier Transforms of Stable Signals

A1·1 Fourier Transform in L^1

This first chapter gives the definition and elementary properties of the Fourier transform of integrable functions, which constitute the specific calculus mentioned in the introduction. Besides linearity, the toolbox of this calculus contains the differentiation rule and the convolution–multiplication rule. The general problem of recovering a function from its Fourier transform then receives a partial answer that will be completed by the results on pointwise convergence of Chapter A3.

We first introduce the notation: $\mathbb{N}$, $\mathbb{Z}$, $\mathbb{Q}$, $\mathbb{R}$, $\mathbb{C}$ are the sets of, respectively, integers, relative integers, rationals, real numbers, complex numbers; $\mathbb{N}_+$ and $\mathbb{R}_+$ are the sets of positive integers and nonnegative real numbers.

In signal theory, functions from $\mathbb{R}$ to $\mathbb{C}$ are called (complex) *signals*. We shall use both terminologies (function, or signal), depending on whether the context is theoretical or applied.

We denote by $L^1_{\mathbb{C}}(\mathbb{R})$ (and sometimes, for short, L^1) the set of functions $f(t)$ [10] from $\mathbb{R}$ into $\mathbb{C}$ such that

$$\int_{\mathbb{R}} |f(t)|\, dt < \infty.$$

In analysis, such functions are called *integrable*. In systems theory, they are called *stable* signals.

---

[10] We shall often use this kind of loose notation, where a phrase such as "the function $f(t)$" means "the function $f : \mathbb{R} \mapsto \mathbb{C}$." We shall also use the notation "f" or "$f(\cdot)$" with a mute argument. For instance, "$f(\cdot - a)$" is the function $t \to f(t - a)$.

Let A be a subset of $\mathbb{R}$. The *indicator function* of A is the function $1_A : \mathbb{R} \mapsto \{0, 1\}$ defined by

$$1_A(t) = \begin{cases} 1 & \text{if } t \in A, \\ 0 & \text{if } t \notin A. \end{cases}$$

The function $f(t)$ is called *locally integrable* if for any closed bounded interval $[a, b] \subset \mathbb{R}$, the function $f(t)1_{[a,b]}(t)$ is integrable. We shall then write

$$f \in L^1_{\mathbb{C},loc}(\mathbb{R})$$

or, for short, $f \in L^1_{loc}$.

The set of functions $f(t)$ from $\mathbb{R}$ into $\mathbb{C}$ such that

$$\int_{\mathbb{R}} |f(t)|^2 < \infty$$

is denoted by $L^2_{\mathbb{C}}(\mathbb{R})$. It is the set of *square-integrable* functions. A signal $f(t)$ in this set is said to have a *finite energy*

$$E = \int_{\mathbb{R}} |f(t)|^2 \, dt.$$

The function $f(t)$ is called *locally square-integrable* if for any closed bounded interval $[a, b] \subset \mathbb{R}$, the function $f(t)1_{[a,b]}(t)$ is square-integrable. We shall then write

$$f \in L^2_{\mathbb{C},loc}(\mathbb{R})$$

or, for short, $f \in L^2_{loc}$.

We recall that in $L^1_{\mathbb{C}}(\mathbb{R})$ or $L^2_{\mathbb{C}}(\mathbb{R})$ two functions are not distinguished if they are equal almost everywhere with respect to the Lebesgue measure.

EXERCISE **A1.1.** *Give an example of a function that is integrable but not of finite energy. Give an example of a function that is of finite energy but not integrable of finite energy. Show that*

$$L^2_{\mathbb{C},loc}(\mathbb{R}) \subset L^1_{\mathbb{C},loc}(\mathbb{R}).$$

A function $f : \mathbb{R} \mapsto \mathbb{C}$ is said to have *bounded support* if there exists a bounded interval $[a, b] \subset \mathbb{R}$ such that $f(t) = 0$ whenever $t \notin [a, b]$.

If the function $f(t)$ is n times continuously differentiable (that is, it has derivatives up to order n, and these derivatives are continuous), we say that it is in C^n. If it is in C^n for all $n \in \mathbb{N}$, it is said to belong to C^∞. The kth derivative of the function $f(t)$, if it exists, is denoted $f^{(k)}(t)$. The 0th derivative is the function itself: $f^{(0)}(t) = f(t)$; in particular, C^0 is the collection of continuous functions from $\mathbb{R}$ to $\mathbb{C}$. The set of continuous functions with bounded support is denoted by C^0_c.

Fourier Transform

We can now give the basic definition.

DEFINITION **A1.1.** *Let $s(t)$ be a stable complex signal. The* Fourier transform *(FT) of $s(t)$ is the function from $\mathbb{R}$ into $\mathbb{C}$:*

$$\hat{s}(\nu) = \int_{\mathbb{R}} s(t)\, e^{-2i\pi\nu t}\, \mathrm{d}t. \tag{1}$$

(Note that the argument of the exponential in the integrand is $-2i\pi\nu t$.) The mapping from the function to its Fourier transform will be denoted by

$$s(t) \stackrel{\mathrm{FT}}{\rightarrow} \hat{s}(\nu) \quad \text{or} \quad \mathcal{F} : s(t) \rightarrow \hat{s}(\nu).$$

Table A1.1 gives the immediate properties of the Fourier transform.

Table A1.1. Elementary Properties of Fourier Transforms

Delay	$s(t - t_0)$	$\stackrel{\mathrm{FT}}{\rightarrow} e^{-2i\pi\nu t_0}\hat{s}(\nu)$
Modulation	$e^{2i\pi\nu_0 t}s(t)$	$\stackrel{\mathrm{FT}}{\rightarrow} \hat{s}(\nu - \nu_0)$
Doppler	$s(at)$	$\stackrel{\mathrm{FT}}{\rightarrow} \frac{1}{\lvert a\rvert}\hat{s}\left(\frac{\nu}{a}\right)$
	$\lambda_1 s_1(t) + \lambda_2 s_2(t)$	$\stackrel{\mathrm{FT}}{\rightarrow} \lambda_1\hat{s}_1(\nu) + \lambda_2\hat{s}_2(\nu)$
	$s^*(t)$	$\stackrel{\mathrm{FT}}{\rightarrow} \hat{s}(-\nu)^*$

EXERCISE **A1.2.** *Prove the assertions in Table A1.1.*

EXERCISE **A1.3.** *Prove the modulation theorem:*

$$s(t)\cos(2\pi\nu_0 t) \stackrel{\mathrm{FT}}{\rightarrow} \frac{1}{2}(\hat{s}(\nu - \nu_0) + \hat{s}(\nu + \nu_0)). \tag{2}$$

(See Fig. A1.1.)

EXERCISE **A1.4.** *Show that the FT of a real signal is* Hermitian even, *that is,*

$$\hat{s}(-\nu) = \hat{s}(\nu)^*.$$

Show that the FT of an odd (resp., even; resp., real and even) signal is odd (resp., even; resp., real and even).

EXERCISE **A1.5.** *Defining the* rectangular pulse

$$\mathrm{rec}_T(t) = 1_{[-\frac{T}{2},+\frac{T}{2}]}(t)$$

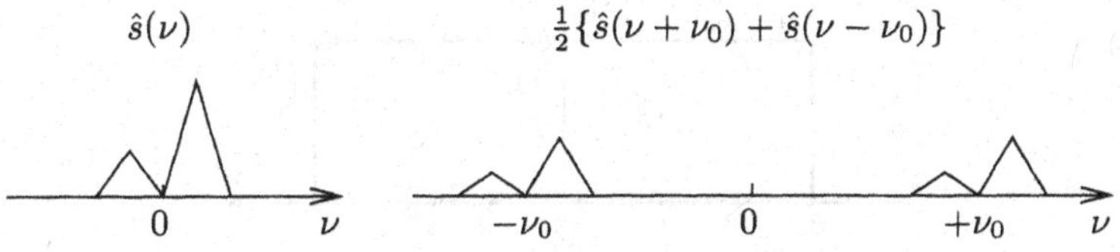

Figure A1.1. Modulation theorem

and the cardinal sine

$$\text{sinc}\,(x) = \frac{\sin(\pi x)}{\pi x},$$

show that (see Fig. A1.2)

$$\text{rec}_T(t) \stackrel{\text{FT}}{\to} T\,\text{sinc}\,(\nu T). \tag{3}$$

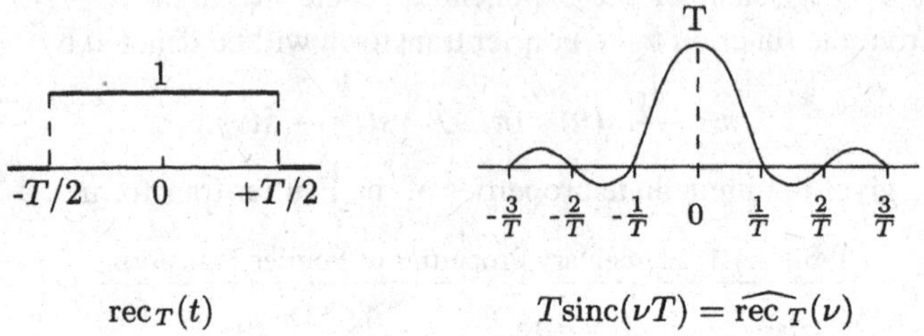

Figure A1.2. Fourier transform of the rectangle function

We will show that the *Gaussian pulse* is its own FT, that is,

$$e^{-\pi t^2} \stackrel{\text{FT}}{\to} e^{-\pi \nu^2}. \tag{4}$$

In order to compute the corresponding Fourier integral, we use contour integration in the complex plane. First, we observe that it is enough to compute the FT $\hat{s}(\nu)$ for $\nu \geq 0$, since this FT is even (see Exercise A1.4). Take $a \geq \nu$ (eventually, a will tend to ∞).

Consider the rectangular contour γ in the complex plane (see Fig. A1.3),

$$\gamma = \gamma_1 + \gamma_2 + \gamma_3 + \gamma_4,$$

where the γ_i's are the oriented line segments

$$\begin{aligned}
\gamma_1 &: (-a, 0) \to (+a, 0),\\
\gamma_2 &: (+a, 0) \to (+a, \nu),\\
\gamma_3 &: (+a, \nu) \to (-a, \nu),\\
\gamma_4 &: (-a, \nu) \to (-a, 0).
\end{aligned}$$

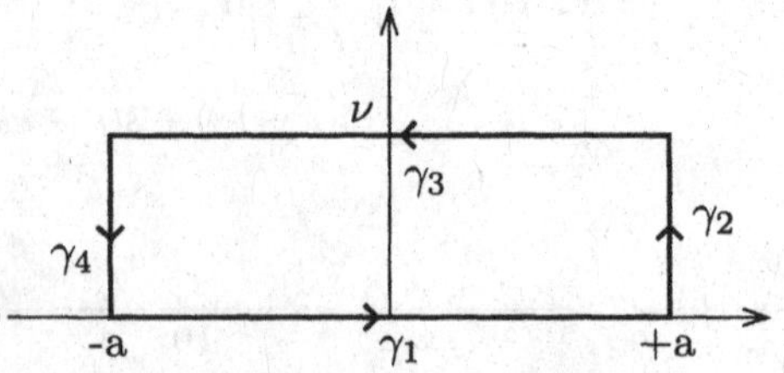

Figure A1.3. The integration path in the proof of (4)

We denote by $-\gamma_i$ the oriented segment whose orientation is opposite that of γ_i. We have

$$\int_\gamma e^{-\pi z^2}\,dz = I_1 + I_2 + I_3 + I_4,$$

where I_i is the integral of $e^{-\pi z^2}$ along γ_i. Since the latter integrand is a holomorphic function, by Cauchy's theorem (see, for instance, Theorem 2.5.2, p. 83, of [A1], or Theorem 2.2, p. 101, of [A6]),

$$\int_\gamma e^{-\pi z^2}\,dz = 0,$$

and therefore,

$$I_1 + I_2 + I_3 + I_4 = 0.$$

We now show that

$$\lim_{a\to\infty} I_2 = \lim_{a\to\infty} I_4 = 0.$$

For I_2, for instance, if we parameterize γ_2 as follows,

$$\gamma_2 = \{a + it; 0 \le t \le v\},$$

then

$$I_2 = \int_0^v e^{-\pi(a+it)^2}\, i\,dt = \int_0^v e^{-\pi(a^2-t^2)}e^{-2i\pi at}\, i\,dt.$$

Therefore, since $v \le a$,

$$|I_2| \le \int_0^a e^{-\pi(a-t)(a+t)}\,dt \le \int_0^a e^{-\pi a(a-t)}\,dt$$

$$= e^{-\pi a^2}\int_0^a e^{\pi at}\,dt = \frac{1}{\pi a}(1 - e^{-\pi a^2}),$$

where the last quantity tends to 0 as a tends to $+\infty$. A similar conclusion holds for I_4, with similar computations. Therefore,

$$\lim_{a\to\infty}(I_1 + I_3) = 0,$$

that is,

$$\lim_{a\to\infty}\int_{\gamma_1} e^{-\pi z^2}\,dz = \lim_{a\to\infty}\int_{-\gamma_3} e^{-\pi z^2}\,dz. \tag{5}$$

Using for γ_1 the obvious parameterization

$$\lim_{a\to\infty}\int_{\gamma_1} = \lim_{a\to\infty}\int_{-a}^{+a} e^{-\pi t^2}\,dt = \int_{\mathbb{R}} e^{-\pi t^2}\,dt = 1.$$

Parameterizing $-\gamma_3$ as follows,

$$-\gamma_3 = \{iv + t; -a \le t \le +a\},$$

we have

$$\int_{-\gamma_3} = \int_{-a}^{+a} e^{-\pi(i\nu+t)^2}\,\mathrm{d}t$$

$$= \int_{-a}^{+a} e^{-\pi(-\nu^2+t^2+2i\nu t)}\,\mathrm{d}t$$

$$= e^{+\pi\nu^2}\int_{-a}^{+a} e^{-\pi t^2} e^{-2i\pi\nu t}\,\mathrm{d}t.$$

Therefore,

$$\lim_{a\to\infty}\int_{-\gamma_3} = e^{-\pi\nu^2}\hat{s}(\nu).$$

Going back to (5), we obtain

$$1 = e^{+\pi\nu^2}\hat{s}(\nu),$$

which gives the announced result. ■

EXERCISE **A1.6.** *Deduce from* (4) *that, for all* $\alpha > 0$,

$$e^{-\alpha t^2} \stackrel{\mathrm{FT}}{\rightarrow} \sqrt{\frac{\pi}{\alpha}}\, e^{-\frac{\pi^2}{\alpha}\nu^2}.$$

The FT of the Gaussian pulse can be obtained by other means (see Exercise A1.16). However, in other cases, contour integration is often necessary.

Using contour integration in the complex plane, we show that, for $a > 0$,

$$s(t) = e^{-at}1_{\mathbb{R}_+}(t) \stackrel{\mathrm{FT}}{\rightarrow} \hat{s}(\nu) = \frac{1}{a+2i\pi\nu}. \tag{6}$$

First observe that

$$\hat{s}(\nu) = \int_0^\infty e^{-2i\pi\nu t}e^{-at}\,\mathrm{d}t = \frac{1}{2i\pi\nu + a}\int_0^\infty e^{-2i\pi\nu t - at}(2i\pi\nu + a)\,\mathrm{d}t$$

$$= \frac{1}{2i\pi\nu + a}\int_\gamma e^{-z}\,\mathrm{d}z.$$

(The reader is refered to Fig. A1.4 for the definition of the lines γ, γ_1, γ_2, and γ_3.)

Therefore, it suffices to show that

$$\int_\gamma e^{-z}\,\mathrm{d}z = 1.$$

By Cauchy's theorem,

$$\int_{\gamma_1} e^{-z}\,\mathrm{d}z + \int_{\gamma_2} e^{-z}\,\mathrm{d}z + \int_{\gamma_3} e^{-z}\,\mathrm{d}z = 0.$$

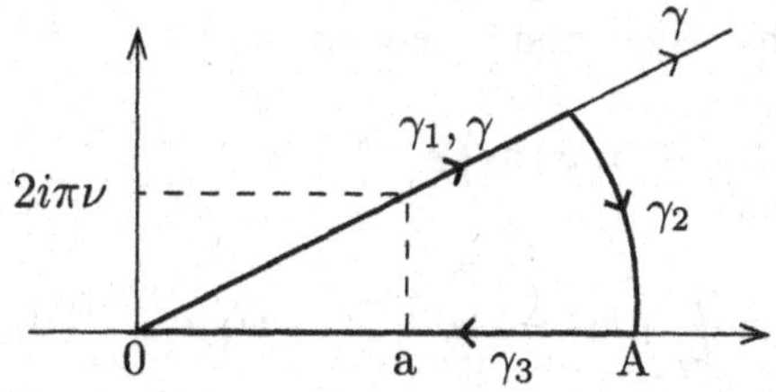

Figure A1.4. The integration path in the proof of (6)

The limit as $A \uparrow \infty$ of $\int_{\gamma_1} e^{-z}\,\mathrm{d}z$ is $\int_{\gamma} e^{-z}\,\mathrm{d}z$, and that of $\int_{\gamma_3} e^{-z}\,\mathrm{d}z = \int_0^A e^{-t}\,\mathrm{d}t$ is 1. It therefore remains to show that the limit as $A \uparrow \infty$ of $\int_{\gamma_2} e^{-z}\,\mathrm{d}z$ is 0, and this follows from the bound

$$\left|\int_{\gamma_2} e^{-z}\,\mathrm{d}z\right| \le e^{-A}|\gamma_2|,$$

where $|\gamma_2| = K \times A$ is the length of γ_2. ■

EXERCISE **A1.7.** *Deduce from* (6) *that*

$$s(t) = e^{-a|t|} \stackrel{\mathrm{FT}}{\to} \hat{s}(\nu) = \frac{2a}{a^2 + 4\pi^2\nu^2}\,.$$

Convolution–Multiplication Rule

THEOREM **A1.1.** *Let $h(t)$ and $x(t)$ be two stable signals. Then the right-hand side of*

$$y(t) = \int_{\mathbb{R}} h(t-s)x(s)\,\mathrm{d}s \tag{7}$$

is defined for almost all t and defines almost everywhere a stable signal whose FT is $\hat{y}(\nu) = \hat{h}(\nu)\hat{x}(\nu)$.

Proof: By Tonelli's theorem and the integrability assumptions,

$$\iint_{\mathbb{R}\times\mathbb{R}} |h(t-s)|\,|x(s)|\,\mathrm{d}t\,\mathrm{d}s = \left(\int_{\mathbb{R}} |h(t)|\,\mathrm{d}t\right)\left(\int_{\mathbb{R}} |x(t)|\,\mathrm{d}t\right) < \infty.$$

This implies that, for almost all t,

$$\int_{\mathbb{R}} |h(t-s)x(s)|\,\mathrm{d}s < \infty.$$

The integral $\int_{\mathbb{R}} h(t-s)x(s)\,\mathrm{d}s$ is therefore well defined for almost all t. Also,

$$\int_{\mathbb{R}} |y(t)|\,\mathrm{d}t = \int_{\mathbb{R}} \left|\int_{\mathbb{R}} h(t-s)x(s)\,\mathrm{d}s\right| \mathrm{d}t$$

$$\le \int_{\mathbb{R}}\int_{\mathbb{R}} |h(t-s)x(s)|\,\mathrm{d}t\,\mathrm{d}s < \infty.$$

Therefore, $y(t)$ is stable. By Fubini's theorem,

$$\begin{aligned}\int_{\mathbb{R}}\left(\int_{\mathbb{R}} h(t-s)x(s)\,\mathrm{d}s\right)e^{-2i\pi\nu t}\,\mathrm{d}t\\ &= \int_{\mathbb{R}}\int_{\mathbb{R}} h(t-s)e^{-2i\pi\nu(t-s)}x(s)e^{-2i\pi\nu s}\,\mathrm{d}s\,\mathrm{d}t\\ &= \int_{\mathbb{R}} x(s)e^{-2i\pi\nu s}\left(\int_{\mathbb{R}} h(t-s)e^{-2i\pi\nu(t-s)}\mathrm{d}t\right)\mathrm{d}s\\ &= \hat{h}(\nu)\hat{x}(\nu).\end{aligned}$$

■

The function $y(t)$, the *convolution* of $h(t)$ with $x(t)$, is denoted by

$$y(t) = (h * x)(t).$$

We therefore have the *convolution–multiplication rule*,

$$(h * x)(t) \xrightarrow{\text{FT}} \hat{h}(\nu)\hat{x}(\nu). \tag{8}$$

EXAMPLE **A1.1.** *The convolution of the rectangular pulse* $\text{rec}_T(t)$ *with itself is the* triangular pulse *of base* $[-T, +T]$ *and height* T,

$$\text{Tri}_T(t) = (T - |t|)1_{[-T,+T]}(t).$$

By the convolution–multiplication rule,

$$\text{Tri}_T(t) \xrightarrow{\text{FT}} (T\,\text{sinc}\,(\nu T))^2 \tag{9}$$

(see Fig. A1.5).

EXERCISE **A1.8.** *Let* $x(t)$ *be a stable complex signal. Show that its* autocorrelation function

$$c(t) = \int_{\mathbb{R}} x(s+t)x^*(s)\,\mathrm{d}s$$

is well defined and integrable and that its FT is $|\hat{x}(\nu)|^2$.

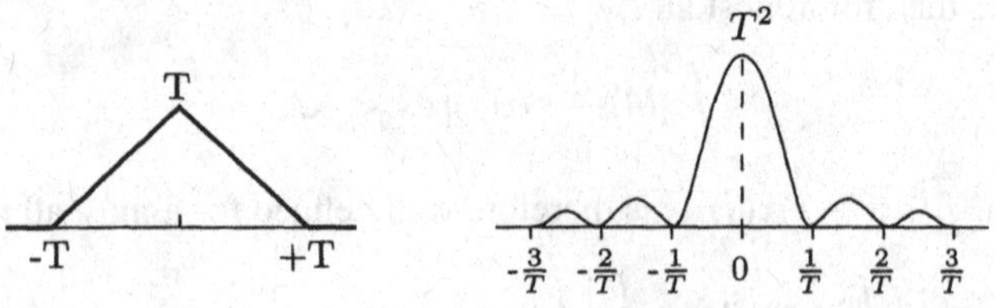

$Tri_T(t)$ $\qquad$ $T^2 sinc^2(\nu T) = \widehat{Tri_T}(\nu)$

Figure A1.5. FT of the triangle function

EXERCISE **A1.9.** *Show that the nth convolution power of* $f(t) = e^{-at}1_{t\geq 0}(t)$, *where* $a > 0$, *is*

$$f^{*n}(t) = \frac{t^{n-1}}{(n-1)!}e^{-at}1_{t\geq 0}(t).$$

($f^{*3} = f * f * f$*, etc.) Deduce from this the FT of* $s(t) = t^n e^{-at}1_{t\geq 0}(t)$.

Riemann–Lebesgue Lemma

The Riemann–Lebesgue lemma[11] is one of the most important technical tools in Fourier analysis, and we shall use it several times, especially in the study of pointwise convergence of Fourier series (Chapter A3).

THEOREM **A1.2.** *The FT of a stable complex signal* $s(t)$ *satisfies*

$$\lim_{|\nu|\to\infty} |\hat{s}(\nu)| = 0. \tag{10}$$

Proof: The FT of a rectangular pulse $s(t)$ satisfies $|\hat{s}(\nu)| \leq K/|\nu|$ [see Eq. (3)]. Hence every signal $s(t)$ that is a finite linear combination of indicator functions of intervals satisfies the same property. Such finite combinations are dense in $L^1_{\mathbb{C}}(\mathbb{R})$ (Theorem 28 of the appendix), and therefore there exists a sequence $s_n(t)$ of integrable functions such that

$$\lim_{n\to\infty} \int_{\mathbb{R}} |s_n(t) - s(t)|\, dt = 0$$

and

$$|\hat{s}_n(\nu)| \leq \frac{K_n}{|\nu|},$$

for finite numbers K_n. From the inequality

$$|\hat{s}(\nu) - \hat{s}_n(\nu)| \leq \int_{\mathbb{R}} |s(t) - s_n(t)|\, dt,$$

we deduce that

$$|\hat{s}(\nu)| \leq |\hat{s}_n(\nu)| + \int_{\mathbb{R}} |s(t) - s_n(t)|\, dt$$

$$\leq \frac{K_n}{|\nu|} + \int_{\mathbb{R}} |s(t) - s_n(t)|\, dt,$$

from which the conclusion follows easily. ■

The following uniform version of the Riemann–Lebesgue lemma will be needed in the sequel.

---

[11] Riemann, B., (1896), Sur la possibilité de représenter une fonction par une série trigonométrique, *Oeuvre Math.*, p. 258.

THEOREM **A1.3.** *Let $f(t)$ be a 2π-periodic locally integrable function, and let $g : [a, b] \mapsto \mathbb{C}$ be in C^1, where $[a, b] \subseteq [-\pi, +\pi]$. Then*

$$\lim_{\lambda\to\infty} \int_a^b f(x-u)g(u)\sin(\lambda u)\,du = 0$$

uniformly in x.

Proof: For arbitrary $\varepsilon > 0$, choose a 2π-periodic function $h(t)$ in C^1 such that

$$\int_{-\pi}^{+\pi} |f(x) - h(x)|\,dx < \varepsilon$$

(Theorem 29 of the appendix). Integrating by parts yields

$$I(\lambda) = \int_a^b h(x-u)g(u)\sin(\lambda u)\,du$$

$$= -\,h(x-u)g(u)\frac{\cos(\lambda u)}{\lambda}\bigg|_a^b + \int_a^b [h(x-u)g(u)]' \frac{\cos(\lambda u)}{\lambda}\,du.$$

Since $h \in C^1$ and is periodic, h and h' are uniformly bounded. The same is true of g, g' (g is in C^1). Therefore,

$$\lim_{\lambda\to\infty} I(\lambda) = 0 \quad \text{uniformly in } x.$$

Now,

$$\left|\int_a^b f(x-u)g(u)\sin(\lambda u)\,du\right|$$

$$\le |I(\lambda)| + \int_a^b |h(x-u) - f(x-u)|\,|g(u)|\sin(\lambda u)\,du$$

$$\le |I(\lambda)| + \max_{a\le u\le b} |g(u)| \int_a^b |h(x-u) - f(x-u)|\sin(\lambda u)\,du$$

$$\le |I(\lambda)| + \max_{a\le u\le b} |g(u)|\varepsilon.$$

The conclusion then follows because ε is arbitrary. ■

A1·2 Inversion Formula

EXERCISE **A1.10.** *Show that the FT of a stable signal is uniformly bounded and uniformly continuous.*

Despite the fact that the FT of an integrable signal is uniformly bounded and uniformly continuous, it is not necessarily integrable. For instance, the FT of the rectangular pulse is the cardinal sine, a non-integrable function. When its FT is integrable, a signal admits a Fourier decomposition.

THEOREM **A1.4.** *Let $s(t)$ be an integrable complex signal with the Fourier transform $\hat{s}(\nu)$. Under the additional condition*

$$\int_{\mathbb{R}} |\hat{s}(\nu)|\, d\nu < \infty, \tag{11}$$

the inversion formula

$$s(t) = \int_{\mathbb{R}} \hat{s}(\nu) e^{+2i\pi\nu t}\, d\nu \tag{12}$$

holds for almost all t. If $s(t)$ is, in addition to the above assumptions, continuous, *equality in* (12) *holds for all t.*

(Note that the exponent of the exponential of the integrand is $+2i\pi\nu t$.)

EXERCISE **A1.11.** *Check that the above result is true for the signal*

$$e_{\alpha,a}(t) = e^{-\alpha t^2 + at} \qquad (\alpha \in \mathbb{R},\ \alpha > 0,\ a \in \mathbb{C}).$$

Proof: We now proceed to the proof of the inversion formula. (It is rather technical and can be skipped in a first reading.) Let $s(t)$ be a stable signal and consider the Gaussian density function

$$h_\sigma(t) = \frac{1}{\sigma\sqrt{2\pi}} e^{-\frac{t^2}{2\sigma^2}}$$

with the FT

$$\hat{h}_\sigma(\nu) = e^{-2\pi^2\sigma^2\nu^2}.$$

We first show that the inversion formula is true for the convolution $(s * h_\sigma)(t)$. Indeed,

$$(s * h_\sigma)(t) = \int_{\mathbb{R}} s(u) h_\sigma(u) e_{\frac{1}{2\sigma^2}, \frac{u}{\sigma^2}}(t)\, du, \tag{13}$$

and the FT of this signal is, by the convolution–multiplication formula, $\hat{s}(\nu)\hat{h}_\sigma(\nu)$. Computing this FT directly from the right-hand side of (13), we obtain

$$\begin{aligned}\hat{s}(\nu)\hat{h}_\sigma(\nu) &= \int_{\mathbb{R}} s(u) h_\sigma(u) \left(\int_{\mathbb{R}} e_{\frac{1}{2\sigma^2}, \frac{u}{\sigma^2}}(t) e^{-2i\pi\nu t}\, dt \right) du \\ &= \int_{\mathbb{R}} s(u) h_\sigma(u) \hat{e}_{\frac{1}{2\sigma^2}, \frac{u}{\sigma^2}}(\nu)\, du.\end{aligned}$$

Therefore, using the result of Exercise A1.11,

$$\begin{aligned}\int_{\mathbb{R}} \hat{s}(\nu)\hat{h}_\sigma(\nu) e^{2i\pi\nu t}\, d\nu &= \int_{\mathbb{R}} \left(\int_{\mathbb{R}} s(u) h_\sigma(u) \hat{e}_{\frac{1}{2\sigma^2}, \frac{u}{\sigma^2}}(\nu)\, du \right) e^{2i\pi\nu t}\, d\nu \\ &= \int_{\mathbb{R}} s(u) h_\sigma(u) e_{\frac{1}{2\sigma^2}, \frac{u}{\sigma^2}}(t)\, du \\ &= (s * h_\sigma)(t).\end{aligned}$$

Thus, we have

$$(s * h_\sigma)(t) = \int_{\mathbb{R}} \hat{s}(\nu)\hat{h}_\sigma(\nu)e^{2i\pi\nu t}\,d\nu, \tag{14}$$

and this is the inversion formula for $(s * h_\sigma)(t)$.

Since for all $\nu \in \mathbb{R}$, $\lim_{\sigma\downarrow 0} \nu \uparrow \hat{h}_\sigma(\nu) = 1$, it follows from Lebesgue's dominated convergence theorem that when $\sigma \downarrow 0$ the right-hand side of (14) tends to

$$\int_{\mathbb{R}} \hat{s}(\nu)e^{2i\pi\nu t}\,d\nu$$

for all $t \in \mathbb{R}$. If we can prove that when $\sigma \downarrow 0$ the function on the left-hand side of (14) converges in $L^1_{\mathbb{C}}(\mathbb{R})$ to the function $s(t)$, then, for almost all $t \in \mathbb{R}$, we have the announced equality (Theorem 25 of the appendix).

To prove convergence in $L^1_{\mathbb{C}}(\mathbb{R})$, we observe that

$$\int_{\mathbb{R}} |(s * h_\sigma)(t) - s(t)|\,dt = \int_{\mathbb{R}} \left| \int_{\mathbb{R}} (s(t-u) - s(t))h_\sigma(u)\,du \right| dt \tag{15}$$

(using the fact $\int_{\mathbb{R}} h_\sigma(u)\,du = 1$), and therefore, defining $f(u) = \int_{\mathbb{R}} |s(t-u) - s(t)|\,dt$,

$$\int_{\mathbb{R}} |s * h_\sigma(t) - s(t)|\,dt \le \int_{\mathbb{R}} f(u)h_\sigma(u)\,du.$$

Now, $|f(u)|$ is bounded (by $2\int_{\mathbb{R}} |s(t)|\,dt$). Therefore, if $\lim_{u\downarrow 0} f(u) = 0$, then, by dominated convergence,

$$\lim_{\sigma\downarrow 0} \int_{\mathbb{R}} f(u)h_\sigma(u) = \lim_{\sigma\downarrow 0} \int_{\mathbb{R}} f(\sigma u)h_1(u)\,du = 0. \tag{16}$$

To prove that $\lim_{u\downarrow 0} f(u) = 0$, we begin with the case where $s(t)$ is continuous with compact support. In particular, it is uniformly bounded. Since we are interested in a limit as u tends to 0, we may suppose that u is in a bounded interval around 0, and in particular, the function $t \to |s(t-u) - s(t)|$ is bounded uniformly in u by an integrable function. It follows from the dominated convergence theorem that $\lim_{u\downarrow 0} f(u) = 0$.

Now, let $s(t)$ be only integrable. Let $\{s_n(\cdot)\}_{n\ge 1}$ be a sequence of continuous functions with compact support that converges in $L^1_{\mathbb{C}}(\mathbb{R})$ to $s(\cdot)$ (Theorem 27 of the appendix). Writing

$$f(u) \le d(s(\cdot - u), s_n(\cdot - u)) + \int_{\mathbb{R}} |s_n(t-u) - s_n(t)|\,dt + d(s(\cdot), s_n(\cdot)),$$

where

$$d(s(\cdot - u), s_n(\cdot - u)) = \int_{\mathbb{R}} |s(t-u) - s_n(t-u)|\,dt,$$

the result easily follows.

Suppose that, in addition, $s(t)$ is continuous. The right-hand side of (12) defines a continuous function because $\hat{s}(\nu)$ is integrable. The everywhere equality in (12) follows from the fact that two continuous functions that are almost everywhere equal are necessarily *everywhere* equal (Theorem 8). ■

The Fourier transform characterizes a stable signal:

COROLLARY **A1.1.** *If two stable signals $s_1(t)$ and $s_2(t)$ have the same Fourier transform, then they are equal almost everywhere.*

Proof: The signal $s(t) = s_1(t) - s_2(t)$ has the FT $\hat{s}(\nu) = 0$, which is integrable, and thus by (12), $s(t) = 0$ for almost all t. ■

EXERCISE **A1.12.** *Give the FT of $s(t) = 1/A(a^2 + t^2)$. Deduce from this the value of the integral*

$$I(t) = \int_{\mathbb{R}} \frac{1}{t^2 + u^2}\, du, \qquad t > 0.$$

EXERCISE **A1.13.** *Deduce from the Fourier inversion formula that*

$$\int_{\mathbb{R}} \left(\frac{\sin(t)}{t}\right)^2 dt = \pi.$$

Exercise 1.14 is very important. It shows that for signals that cannot be called pathological, the version of the Fourier inversion theorem that we have in this chapter is not applicable, and therefore we shall need finer results, which are given in Chapter A3.

EXERCISE **A1.14.** *Let $s(t)$ be a stable right-continuous signal, with a limit from the left at all times. Show that if $s(t)$ is discontinuous at some time t_0, its FT cannot be integrable.*

Regularization Lemma

In the course of the proof of Theorem A1.4, we have used a special case of the *regularization lemma* below, which is very useful in many circumstances.

DEFINITION **A1.2.** *A regularizing function is a nonnegative function $h_\sigma : \mathbb{R} \to \mathbb{R}$ depending on a parameter $\sigma > 0$ and such that*

$$\int_{\mathbb{R}} h_\sigma(u)\, du = 1, \qquad \textit{for all } \sigma > 0,$$

$$\lim_{\sigma \downarrow 0} \int_{-a}^{+a} h_\sigma(u)\, du = 1, \qquad \textit{for all } a > 0,$$

$$\lim_{\sigma \downarrow 0} h_\sigma(u) = 1, \qquad \textit{for all } u \in \mathbb{R}.$$

LEMMA **A1.1.** *Let $h_\sigma : \mathbb{R} \to \mathbb{R}$ be a regularizing function. Let $s(t)$ be in $L^1_{\mathbb{C}}(\mathbb{R})$. Then*

$$\lim_{\sigma \downarrow 0} \int_{\mathbb{R}} |(s * h_\sigma)(t) - s(t)|\, dt = 0.$$

Proof: We can use the proof of Theorem A1.4, starting from (15). The only place where the specific form of h_σ (a Gaussian density) is used is (16). We must therefore prove that

$$\lim_{\sigma \downarrow 0} \int_{\mathbb{R}} f(u) h_\sigma(u) = 0$$

independently. Fix $\varepsilon > 0$. Since $\lim_{u \downarrow 0} f(u) = 0$, there exists $a = a(\varepsilon)$ such that

$$\int_{-a}^{+a} f(u) h_\sigma(u)\, du \leq \frac{\varepsilon}{2} \int_{-a}^{+a} h_\sigma(u)\, du \leq \frac{\varepsilon}{2}.$$

Since $f(u)$ is bounded (say, by M),

$$\int_{\mathbb{R}\setminus[-a,+a]} f(u) h_\sigma(u)\, du \leq M \int_{\mathbb{R}\setminus[-a,+a]} h_\sigma(u)\, du.$$

The last integral is, for sufficiently small σ, less than $\varepsilon/2M$. Therefore, for sufficiently small σ,

$$\int_{\mathbb{R}} f(u) h_\sigma(u)\, du \leq \frac{\varepsilon}{2} + \frac{\varepsilon}{2} = \varepsilon. \qquad \blacksquare$$

The function h_σ is an approximation of the Dirac generalized function $\delta(t)$ in that, for all $\varphi \in C_c^0$,

$$\lim_{\sigma \downarrow 0} \int_{\mathbb{R}} h_\sigma(t)\, \varphi(t)\, dt = \varphi(0) = \int_{\mathbb{R}} \delta(t)\, \varphi(t)\, dt.$$

The last equality is symbolic and defines the Dirac generalized function (see Section B2·4). The first equality is obtained as in the proof of the previous lemma, this time letting $f(u) = \varphi(u) - \varphi(0)$.

Differentiation in the Frequency Domain

We shall see how differentiation in the time domain is expressed in the frequency domain.

THEOREM **A1.5.** *(a) If the integrable signal $s(t)$ is such that $t^k s(t) \in L^1_{\mathbb{C}}(\mathbb{R})$ for all $1 \leq k \leq n$, then its FT is in C^n, and*

$$(-2i\pi t)^k s(t) \xrightarrow{\text{FT}} \hat{s}^{(k)}(\nu) \quad \text{for all } 1 \leq k \leq n.$$

(b) If the signal $s(t) \in C^n$ and if it is, together with its n first derivatives, integrable, then

$$s^{(k)}(t) \xrightarrow{\text{FT}} (2i\pi\nu)^k \hat{s}(\nu) \quad \text{for all } 1 \leq k \leq n.$$

Proof: (a) In the right-hand side of the expression

$$\hat{s}(\nu) = \int_{\mathbb{R}} e^{-2i\pi\nu t} s(t)\, dt,$$

we can differentiate k times under the integral sign (see Theorem 15 and the hypothesis $t^k s(t) \in L^1_{\mathbb{C}}(\mathbb{R})$) and obtain

$$\hat{s}^{(k)}(\nu) = \int_{\mathbb{R}} (-2i\pi t)^k e^{-2i\pi\nu t} s(t)\, dt.$$

(b) It suffices to prove this for $n = 1$, and iterate the result. We first observe that $\lim_{|a|\uparrow\infty} s(a) = 0$. Indeed, with $a > 0$, for instance,

$$s(a) = s(0) + \int_0^a s'(t)\, dt,$$

and therefore, since $s'(t) \in L^1_{\mathbb{C}}(\mathbb{R})$, the limit exists and is finite. This limit must be 0 because $s(t)$ is integrable. Now, the FT of $s'(t)$ is

$$\int_{\mathbb{R}} e^{-2i\pi\nu t} s'(t)\, dt = \lim_{a\uparrow\infty} \int_{-a}^{+a} e^{-2i\pi\nu t} s'(t)\, dt.$$

Integration by parts yields

$$\int_{-a}^{+a} e^{-2i\pi\nu t} s'(t)\, dt = \left(e^{-2i\pi\nu t} s(t)\right)_{-a}^{+a} + \int_{-a}^{+a} (2i\pi\nu) e^{-2i\pi\nu t} s(t)\, dt.$$

It then suffices to let a tend to ∞ to obtain the announced result. ■

EXERCISE **A1.15.** *Let $s(t)$ be a stable signal with a Fourier transform with compact support. Show that $s(t) \in C^\infty$, that all its derivatives are integrable, and that the kth derivative has the FT $(2i\pi\nu)^k \hat{s}(\nu)$.*

EXERCISE **A1.16.** *Give a differential equation satisfied by the Gaussian pulse, and use it to deduce its Fourier transform. Could you do the same to prove* (6)*?*

The Heat Equation

We now pay our tribute to the founder and give the solution of the heat equation, which was announced in the introduction. Recall that the heat equation relative to an infinite rod is the partial differential equation

$$\frac{\partial\theta}{\partial t} = \kappa \frac{\partial^2\theta}{\partial^2 x}, \tag{17}$$

where $\theta(x, t)$ is the temperature at time t and at location x of the rod with heat conductance κ, and with the given initial temperature distribution

$$\theta(x, 0) = f(x). \tag{18}$$

We assume that f is integrable.

Let $\xi \mapsto \Theta(\xi, t)$ be the FT of $x \mapsto \theta(x, t)$. (We take different notations because the variable with respect to which the FT is taken is not the time variable t but the space variable x.) In the Fourier domain, Eq. (17) becomes

$$\frac{d\,\Theta(\xi, t)}{dt} = -\kappa(4\pi^2\xi^2)\Theta(\xi, t),$$

with the initial condition

$$\Theta(\xi, 0) = F(\xi),$$

where $F(\xi)$ is the FT of $f(x)$. The solution is

$$\Theta(\xi, t) = F(\xi)e^{-4\pi^2\kappa\xi^2 t}.$$

Since $x \mapsto (4\pi\kappa t)^{-1/2}e^{(4\kappa t)^{-1/2}x^2}$ has the FT $\xi \mapsto e^{-4\pi^2\kappa\xi^2 t}$, the convolution–multiplication formula gives

$$\theta(x, t) = (4\pi\kappa t)^{-\frac{1}{2}} \int_{\mathbb{R}} f(x - y)e^{-\frac{y^2}{4\kappa t}} \, dy,$$

or

$$\theta(x, t) = \frac{1}{\sqrt{\pi}} \int_{\mathbb{R}} f(x - 2\sqrt{\kappa t}y)e^{-y^2} \, dy. \blacksquare$$

As we mentioned earlier, Fourier considered the finite rod heat equation, which receives a similar solution, in terms of Fourier series rather than Fourier integrals (see Chapter A2). The efficiency of the Fourier method in solving differential or partial differential equations of mathematical physics has been, after the pioneering work of Fourier, amply demonstrated[12].

---

[12] See, for instance, the classic text of I. N. Sneddon, *Fourier Transforms*, McGraw–Hill, 1951; Dover edition, 1995.

A2

Fourier Series of Locally Stable Periodic Signals

A2·1 Fourier Series in L^1_{loc}

Fourier Coefficients

A periodic signal is neither stable nor of finite energy unless it is almost everywhere null, and therefore, the theory of the preceding Chapter is not applicable. The relevant notion is that of Fourier series. (Note that Fourier series were introduced before Fourier transforms, in contrast with the order of appearance chosen in this text.) The elementary theory of Fourier series of this section is parallel to the elementary theory of Fourier transforms of the previous section. The connection between Fourier transforms and Fourier series is made by the Poisson sum formula, of which we present a weak (yet useful) version in this chapter.

A complex signal $s(t)$ is called *periodic* with period $T > 0$ (or T-periodic) if, for all $t \in \mathbb{R}$,

$$s(t+T) = s(t).$$

A T-periodic signal $s(t)$ is *locally stable*, or *locally integrable*, if $s(t) \in L^1_{\mathbb{C}}([0, T])$, that is,

$$\int_0^T |s(t)|\, dt < \infty.$$

A T-periodic signal $s(t)$ is *locally square-integrable* if $s(t) \in L^2_{\mathbb{C}}([0, T])$, that is,

$$\int_0^T |s(t)|^2\, dt < \infty.$$

One also says in this case that $s(t)$ has *finite power*, since

$$\lim_{A\to\infty} \frac{1}{A}\int_0^A |s(t)|^2 = \frac{1}{T}\int_0^T |s(t)|^2 \, dt < \infty.$$

As the Lebesgue measure of $[0, T]$ is finite, $L^2_{\mathbb{C}}([0,T]) \subset L^1_{\mathbb{C}}([0,T])$. (See Theorem 19 of the appendix.) In particular, a finite-power periodic signal is also locally stable.

We are now ready for the basic definition.

DEFINITION **A2.1.** *The Fourier transform* $\{\hat{s}_n\}$, $n \in \mathbb{Z}$, *of the locally stable* T-*periodic signal* $s(t)$ *is defined by the formula*

$$\hat{s}_n = \frac{1}{T}\int_0^T s(t) e^{-2i\pi \frac{n}{T} t} \, dt, \tag{19}$$

and $\hat{s}_n$ *is called the nth* Fourier coefficient *of the signal* $s(t)$.

EXERCISE **A2.1.** *Compute the Fourier coefficients of the* T*-periodic function* $s(t)$ *such that on* $[0, T)$, $s(t) = t$.

EXERCISE **A2.2.** *Let* $s(t)$ *be a locally stable* T*-periodic signal. Defining*

$$s_T(t) = s(t) 1_{[0,T]}(t),$$

show that the nth Fourier coefficient $\hat{s}_n$ *of* $s(t)$ *and the FT* $\widehat{s_T}(\nu)$ *of* $s_T(t)$ *are linked by*

$$\hat{s}_n = \frac{1}{T}\widehat{s_T}\left(\frac{n}{T}\right). \tag{20}$$

EXERCISE **A2.3.** *Compute the Fourier coefficients of the* T*-periodic signal* $s(t)$ *such that on* $[-T/2, +T/2)$, $s(t) = 1_{[-\alpha\frac{T}{2}, +\alpha\frac{T}{2}]}(t)$, *where* $\alpha \in (0, 1)$.

EXERCISE **A2.4.** *Let* $s(t)$ *be a* T*-periodic locally stable signal with nth Fourier coefficient* $\hat{s}_n$. *Show that* $\lim_{|n|\uparrow\infty} \hat{s}_n = 0$.

One often represents the sequence $\{\hat{s}_n\}_{n\in\mathbb{Z}}$ of the Fourier coefficients of a T-periodic signal by "spectral lines" separated by $1/T$ from each other along the frequency axis. The spectral line at frequency n/T has the complex amplitude $\hat{s}_n$ (see Fig. A2.1). This is sometimes interpreted by saying that the FT of $s(t)$ is

$$\hat{s}(\nu) = \sum_{n\in\mathbb{Z}} \hat{s}_n \delta\left(\nu - \frac{n}{T}\right),$$

where $\delta(t)$ is the Dirac generalized function (see Section B2·4).

EXERCISE **A2.5.** *Let* $s(t)$ *be a* T*-periodic locally stable signal with nth Fourier coefficient* $\hat{s}_n$. *What is the nth Fourier coefficient of* $s(t - a)$, *where* $a \in \mathbb{R}$? *What can you say about the period and the Fourier coefficients of the signal* $s(t/a)$, *where* $a > 0$?

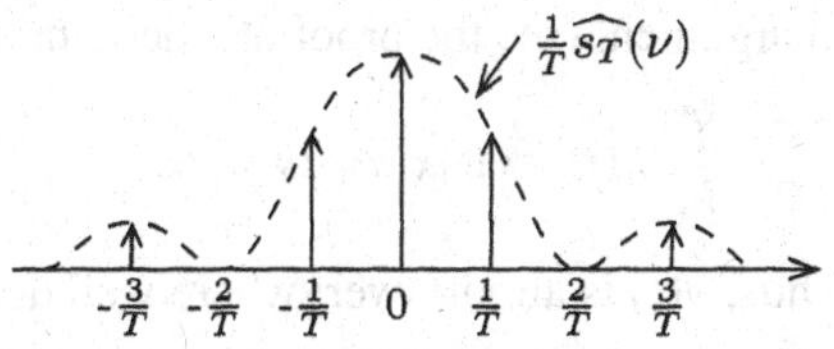

Figure A2.1. From the Fourier transform to the Fourier coefficients

Convolution–Multiplication Rule

THEOREM **A2.1.** *Let $x(t)$ be a T-periodic locally stable signal, and let $h(t)$ be a stable signal. The signal*

$$y(t) = \int_{\mathbb{R}} h(t-s)x(s)\,\mathrm{d}s \tag{21}$$

is almost everywhere well defined, T-periodic, and locally stable. Its nth Fourier coefficient is

$$\hat{y}_n = \hat{h}\left(\frac{n}{T}\right)\hat{x}_n, \tag{22}$$

where $\hat{h}(\nu)$ is the FT of $h(t)$ (see Fig A2.2).

Proof: We have

$$\int_{\mathbb{R}} |h(t-s)|\,|x(s)|\,\mathrm{d}s = \int_0^T |\tilde{h}_T(t-s)|\,|x(s)|\,\mathrm{d}s,$$

where

$$\tilde{h}_T(u) = \sum_{n\in\mathbb{Z}} h(u+nT).$$

Now

$$\int_0^T |\tilde{h}_T(u)|\,\mathrm{d}u \leq \int_{\mathbb{R}} |h(u)|\,\mathrm{d}u < \infty,$$

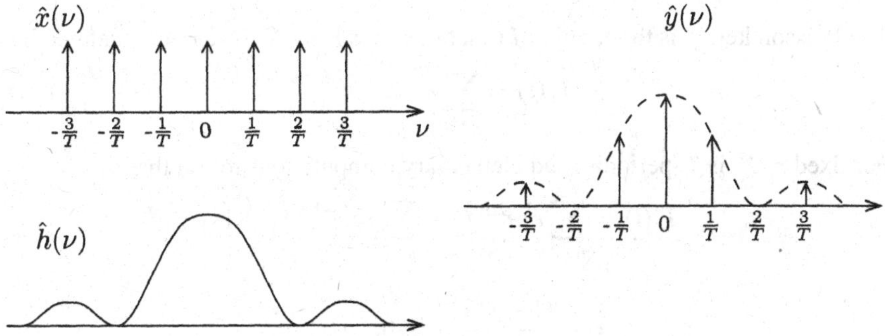

Figure A2.2. Filtering a periodic signal

and hence by the usual argument (see the proof of Theorem A1.1),

$$\int_{\mathbb{R}} |h(t-s)|\,|x(s)|\,\mathrm{d}s < \infty,$$

for almost all $t \in \mathbb{R}$. Thus, $y(t)$ is almost everywhere well defined by (21). Also,

$$y(t+T) = \int_{\mathbb{R}} x(t+T-s)h(s)\,\mathrm{d}s$$

$$= \int_{\mathbb{R}} x(t-s)h(s)\,\mathrm{d}s,$$

which shows that $y(t)$ is periodic with period T. The same argument as in the proof of Theorem A1.1 shows that $y(t)$ is locally stable. Finally,

$$\hat{y}_n = \frac{1}{T}\int_0^T y(t)e^{-2i\pi\frac{n}{T}t}\,\mathrm{d}t$$

$$= \frac{1}{T}\int_0^T\int_0^T \tilde{h}_T(t-s)x(s)e^{-2i\pi\frac{n}{T}t}\,\mathrm{d}t\,\mathrm{d}s$$

$$= \frac{1}{T}\int_0^T\left(\int_0^T \tilde{h}_T(t-s)e^{-2i\pi\frac{n}{T}(t-s)}\,\mathrm{d}s\right)x(s)e^{-2i\pi\frac{n}{T}s}\,\mathrm{d}s$$

$$= \hat{h}\left(\frac{n}{T}\right)\hat{x}_n.$$

■

A2·2 Inversion Formula

The Poisson Kernel

In the proof of the Fourier series inversion formula, the Poisson kernel will play a role similar to that of the Gaussian pulse in the proof of the Fourier transform inversion formula of the previous section.

The Poisson kernel is the family of functions $P_r : \mathbb{R} \mapsto \mathbb{C}$, $0 < r < 1$, defined by

$$P_r(t) = \sum_{n\in\mathbb{Z}} r^{|n|}e^{2i\pi\frac{n}{T}t}. \tag{23}$$

For fixed r, P_r is T-periodic, and elementary computations reveal that

$$P_r(t) = \sum_{n\geq 0} r^n e^{2i\pi\frac{n}{T}t} + \sum_{n\geq 0} r^n e^{-2i\pi\frac{n}{T}t} - 1$$

$$= \frac{(1-r^2)}{\left|1-re^{2i\pi\frac{t}{T}}\right|^2} \geq 0,$$

and therefore,

$$P_r(t) \geq 0. \tag{24}$$

Also,

$$\frac{1}{T}\int_{-T/2}^{+T/2} P_r(t)\,dt = 1. \tag{25}$$

In view of the above expression of the Poisson kernel, we have the bound

$$\frac{1}{T}\int_{[-\frac{T}{2},+\frac{T}{2}]\setminus[-\varepsilon,+\varepsilon]} P_r(t)\,dt \leq \frac{(1-r^2)}{\left|1-e^{2i\pi\frac{\varepsilon}{T}}\right|^2},$$

and therefore, for all $\varepsilon > 0$,

$$\lim_{r\uparrow 1}\frac{1}{T}\int_{[-\frac{T}{2},+\frac{T}{2}]\setminus[-\varepsilon,+\varepsilon]} P_r(t)\,dt = 0. \tag{26}$$

Properties (24)–(25) make of the Poisson kernel a regularizing kernel, and in particular,

$$\lim_{r\uparrow 1}\frac{1}{T}\int_{-\frac{T}{2}}^{+\frac{T}{2}} \varphi(t)P_r(t)\,dt = \varphi(0),$$

for all bounded, continuous $\varphi : \mathbb{R} \to \mathbb{C}$ (same proof as in Lemma A1.1).

The following result is similar to the Fourier inversion formula for stable signals (Theorem A1.4).

THEOREM **A2.2.** *Let $s(t)$ be a T-periodic locally stable complex signal with Fourier coefficients $\{\hat{s}_n\}$, $n \in \mathbb{Z}$. If*

$$\sum_{n\in\mathbb{Z}} |\hat{s}_n| < \infty, \tag{27}$$

then, for almost all $t \in \mathbb{R}$,

$$s(t) = \sum_{n\in\mathbb{Z}} \hat{s}_n e^{+2i\pi\frac{n}{T}t}. \tag{28}$$

If we add to the above hypotheses the assumption that $s(t)$ is a continuous function, then the inversion formula (28) *holds for* all t.

Proof: The proof is similar to that of Theorem A1.4. We have

$$\sum_{n\in\mathbb{Z}} \hat{s}_n r^{|n|} e^{2i\pi\frac{n}{T}t} = \frac{1}{T}\int_{-\frac{T}{2}}^{+\frac{T}{2}} s(u)P_r(t-u)\,du, \tag{29}$$

and

$$\lim_{r\uparrow 1}\int_0^T \left|\int_0^T s(u)P_r(t-u)\,\frac{du}{T} - s(t)\right| dt = 0,$$

that is: The right-hand side of (29) tends to $s(t)$ in $L^1_{\mathbb{C}}([0,T])$ when $r \uparrow 1$. Since $\sum_{n\in\mathbb{Z}} |\hat{s}_n| < \infty$, the function of t in the left-hand side of (29) tends toward the function $\sum_{n\in\mathbb{Z}} \hat{s}_n e^{+2i\pi(n/T)t}$, pointwise and in $L^1_{\mathbb{C}}([0,T])$. The result then follows from Theorem 25.

The statement in the case where $s(t)$ is continuous is proved exactly as the corresponding statement in Theorem A1.4. ■

As in the case of stable signals, we deduce from the inversion formula the *uniqueness theorem.*

COROLLARY **A2.1.** *Two locally stable periodic signals with the same period T that have the same Fourier coefficients are equal almost everywhere.*

EXERCISE **A2.6.** *Compute*

$$\sum_{n\geq 1} 1/n^2$$

using the expression of the Fourier coefficients of the 2-periodic signal $s(t)$ such that

$$s(t) = \text{Tri}_1(t) \qquad \textit{for } t \in [-1,+1].$$

EXERCISE **A2.7.** *Let $x(t)$ be a T-periodic locally stable signal with nth Fourier coefficient $\hat{x}_n$ such that*

$$\sum_{n\in\mathbb{Z}} |n|^p |\hat{x}_n| < \infty.$$

Show that $x(t)$ is p times differentiable and that if the pth derivative is locally integrable, its nth Fourier coefficient is $\left(2i\pi \frac{n}{T}\right)^p \hat{x}_n$.

The Weak Poisson Formula

The Poisson sum formula takes many forms. The strong version is

$$T\sum_{n\in\mathbb{Z}} s(nT) = \sum_{n\in\mathbb{Z}} \hat{s}\left(\frac{n}{T}\right). \tag{30}$$

This aesthetic formula has a number of applications in signal processing (see Part B).

The next result establishes the connection between the Fourier transform and Fourier series, and is central to sampling theory. It is a weak form of the Poisson sum formula (see the discussion after the statement of the theorem).

THEOREM **A2.3.** *Let $s(t)$ be a stable complex signal, and let $0 < T < \infty$ be fixed. The series $\sum_{n\in\mathbb{Z}} s(t+nT)$ converges absolutely almost everywhere to a T-periodic locally integrable function $\Phi(t)$, the nth Fourier coefficient of which is $(1/T)\hat{s}(n/T)$.*

We paraphrase this result as follows: Under the above conditions, the function

$$\Phi(t) := \sum_{n\in\mathbb{Z}} s(t+nT) \tag{31}$$

is T-periodic and locally integrable, and its formal Fourier series is

$$S_f(t) = \frac{1}{T}\sum_{n\in\mathbb{Z}} \hat{s}\left(\frac{n}{T}\right) e^{2i\pi\frac{n}{T}t}. \tag{32}$$

(We speak of a "formal" Fourier series, because nothing is said about its convergence.) Therefore, whenever we are able to show that the Fourier series represents the function at $t = 0$, that is, if $\Phi(0) = S_f(0)$, then we obtain the Poisson sum formula (30).

For now, we are saying nothing about the convergence of the Fourier series. This is why we talk about a *weak* Poisson's formula. A *strong* Poisson's formula corresponds to the case where one can prove the equality *everywhere* (and in particular at $t = 0$) of $\Phi(t)$ and of its Fourier series. We shall say more about the Poisson formula and, in particular, give strong versions of it in Section A3·3. The version we have here, and that we shall proceed to prove, is the one we need in the Shannon–Nyquist sampling theorem (Chapter B2).

Proof: We first show that $\Phi(t)$ is well defined:

$$\begin{aligned}\int_0^T \sum_{n\in\mathbb{Z}} |s(t+nT)|\,\mathrm{d}t &= \sum_{n\in\mathbb{Z}} \int_0^T |s(t+nT)|\,\mathrm{d}t \\ &= \sum_{n\in\mathbb{Z}} \int_{nT}^{(n+1)T} |s(t)|\,\mathrm{d}t = \int_{\mathbb{R}} |s(t)|\,\mathrm{d}t < \infty.\end{aligned}$$

In particular,

$$\sum_{n\in\mathbb{Z}} |s(t+nT)| < \infty \quad \text{a.e.}$$

Therefore, the series $\sum_{n\in\mathbb{Z}} s(t+nT)$ converges absolutely for almost all t. In particular, $\Phi(t)$ is well defined (define it arbitrarily when the series does not converge). This function is clearly T-periodic. We have

$$\begin{aligned}\int_0^T |\Phi(t)|\,\mathrm{d}t &= \int_0^T \left|\sum_{n\in\mathbb{Z}} s(t+nT)\right| \mathrm{d}t \\ &\le \int_0^T \sum_{n\in\mathbb{Z}} |s(t+nT)|\,\mathrm{d}t = \int_{\mathbb{R}} |s(t)|\,\mathrm{d}t < \infty.\end{aligned}$$

Therefore, $\Phi(t)$ is stable. Its nth Fourier coefficient is

$$\begin{aligned}c_n(\Phi) &= \frac{1}{T}\int_0^T \Phi(t) e^{-2i\pi\frac{n}{T}t}\,\mathrm{d}t \\ &= \frac{1}{T}\int_0^T \left\{\sum_{k\in\mathbb{Z}} s(t+kT)\right\} e^{-2i\pi\frac{n}{T}t}\,\mathrm{d}t\end{aligned}$$

$$= \frac{1}{T}\int_0^T \left\{ \sum_{k\in\mathbb{Z}} s(t+kT)e^{-2i\pi\frac{n}{T}(t+kT)} \right\} \mathrm{d}t$$

$$= \frac{1}{T}\int_{\mathbb{R}} s(t)e^{-2i\pi\frac{n}{T}t}\,\mathrm{d}t = \frac{1}{T}\hat{s}\left(\frac{n}{T}\right). \qquad \blacksquare$$

We have a function as well as its formal Fourier series. When both are equal everywhere, we obtain the strong Poisson sum formula. The next exercise gives conditions for this. It will be improved by Theorem A3.12.

EXERCISE **A2.8.** *Let $s(t)$ be a stable signal with the FT $\hat{s}(\nu)$, and suppose that*

(a) $\sum_{n\in\mathbb{Z}} s(t+nT)$ is a continuous function, and

(b) $\sum_{n\in\mathbb{Z}} |\hat{s}(n/T)| < \infty$.

Show that, for all $t \in \mathbb{R}$,

$$\sum_{n\in\mathbb{Z}} s(t+nT) = \sum_{n\in\mathbb{Z}} \hat{s}\left(\frac{n}{T}\right) e^{2i\pi\frac{n}{T}t}.$$

A3

Pointwise Convergence of Fourier Series

A3·1 Dini's and Jordan's Theorems

The inversion formula for Fourier series obtained in Chapter A2 requires a rather strong condition of summability of the Fourier coefficients series. Moreover, this condition implies that the function itself is almost everywhere equal to a continuous function. In this section, the class of functions for which the inversion formula holds is extended.

Recall Kolmogorov's negative result (see the Introduction):

THEOREM **A3.1.** *There exists a locally integrable* 2π*-periodic function* $f : \mathbb{R} \to \mathbb{C}$ *for which the Fourier series diverges everywhere.*

This result challenges one to obtain conditions that a locally integrable 2π-periodic function f must satisfy in order for its Fourier series to converge to f. Recall that the Fourier series associated with a 2π-periodic locally integrable function f is the *formal Fourier series*

$$\sum_{n\in\mathbb{Z}} c_n(f)\, e^{inx}, \tag{33}$$

where $c_n(f)$ is the nth Fourier coefficient

$$c_n(f) = \frac{1}{2\pi} \int_{-\pi}^{+\pi} f(u) e^{-inu}\, du. \tag{34}$$

The series (33) is called *formal* as long as one does not say something about its convergence in some sense (pointwise, almost everywhere, in L^1, etc). If one has no more than the condition that f is 2π-periodic and locally integrable, the worst can happen, as Kolmogorov's theorem shows.

The purpose of this section is to find reasonable conditions guaranteeing convergence as $n \to \infty$ of the truncated Fourier series

$$S_n^f(x) = \sum_{k=-n}^{+n} c_k(f)e^{ikx}. \tag{35}$$

We have to specify (1) in what sense this convergence takes place and (2) what the limit is. Ideally, the convergence should be pointwise and to f itself. The next exercise gives a simple instance where this is true.

EXERCISE **A3.1.** *Assume that the trigonometric series*

$$S_n(t) = \sum_{k=-n}^{+n} c_k e^{ikt}$$

converges uniformly to some function $f(t)$. Show that in this case, for all $k \in \mathbb{Z}$,

$$c_k = c_k(f).$$

Dirichlet's Integral

We will first express the truncated series S_n^f in a form suitable for analysis. For this we write

$$S_n^f(x) = \sum_{k=-n}^{+n} \left\{ \frac{1}{2\pi} \int_{-\pi}^{+\pi} f(s)e^{-iks}\,\mathrm{d}s \right\} e^{ikx}$$

$$= \frac{1}{2\pi} \int_{-\pi}^{+\pi} \left\{ \sum_{k=-n}^{+n} e^{ik(x-s)} \right\} f(s)\,\mathrm{d}s.$$

Elementary computations give

$$\sum_{k=-n}^{+n} e^{ikt} = \frac{\sin((n+\frac{1}{2})t)}{\sin(t/2)}, \tag{36}$$

(the function in the right-hand side is called the *Dirichlet kernel*) and therefore,

$$S_n^f(x) = \frac{1}{2\pi} \int_{-\pi}^{+\pi} \frac{\sin((n+\frac{1}{2})(x-s))}{\sin((x-s)/2)} f(s)\,\mathrm{d}s.$$

Performing the change of variable $x - s = u$ and taking into account the fact that f and the Dirichlet kernel are 2π-periodic, we obtain

$$S_n^f(x) = \frac{1}{2\pi} \int_{-\pi}^{+\pi} \frac{\sin((n+\frac{1}{2})u)}{\sin(u/2)} f(x+u)\,\mathrm{d}u. \tag{37}$$

The right-hand side of (37) is called the *Dirichlet integral* .

If we let $f(t) \equiv 1$ in (35), we obtain 1; on substituting this in (37),

$$\frac{1}{2\pi}\int_{-\pi}^{+\pi}\frac{\sin((n+\frac{1}{2})u)}{\sin(u/2)}\,du = 1. \tag{38}$$

Therefore, for any real number A,

$$S_n^f(x) - A = \frac{1}{2\pi}\int_{-\pi}^{+\pi}\frac{\sin((n+\frac{1}{2})u)}{\sin(u/2)}(f(x+u)-A)\,du \tag{39}$$

or, equivalently,

$$S_n^f(x) - A = \frac{1}{2\pi}\int_{0}^{\pi}\frac{\sin((n+\frac{1}{2})u)}{\sin(u/2)}\{f(x+u)+f(x-u)-2A\}\,du. \tag{40}$$

Therefore, in order to show that, for a given $x \in \mathbb{R}$, $S_n^f(x)$ tends to A as $n \to \infty$, it is necessary and sufficient to show that the Dirichlet integral in the right-hand side of (39) converges to zero as $n \to \infty$.

The *localization principle* states that the convergence of the Fourier series is a local property. More precisely:

THEOREM **A3.2.** *If f and g are two locally integrable 2π-periodic complex-valued functions such that, for a given $x \in \mathbb{R}$ and some $\delta > 0$, it holds that $f(t) = g(t)$ whenever $t \in [x-\delta, x+\delta]$, then*

$$\lim_{n\uparrow\infty}\{S_n^f(x) - S_n^g(x)\} = 0.$$

Proof: Using (39) we have

$$\begin{aligned} S_n^f(x) - S_n^g(x) &= \frac{1}{2\pi}\int_{-\pi}^{+\pi}\sin((n+\tfrac{1}{2})u)\,1_{|u|\geq\delta}\,\frac{f(x+u)-g(x+u)}{\sin(u/2)}\,du \\ &= \frac{1}{2\pi}\int_{-\pi}^{+\pi}\sin((n+\tfrac{1}{2})u)\,w(u)\,du, \end{aligned}$$

where

$$w(u) = 1_{|u|\geq\delta}\,\frac{f(x+u)-g(x+u)}{\sin(u/2)}$$

is integrable over $[0, 2\pi]$. The last integral therefore tends to zero as $n \to \infty$ by the Riemann–Lebesgue lemma. ■

We now state the general *pointwise convergence theorem.*

THEOREM **A3.3.** *Let f be a locally integrable 2π-periodic complex-valued function, and let $x \in \mathbb{R}$ and $A \in \mathbb{R}$ be given. Then*

$$\lim_{n\uparrow\infty} S_n^f(x) = A$$

if, for some $0 < \delta \le \pi$,

$$\lim_{n\uparrow\infty} \int_0^\delta \sin((n+\tfrac{1}{2})u)\,\frac{\phi(u)}{u/2}\,du = 0, \tag{41}$$

where

$$\phi(u) = f(x+u) + f(x-u) - 2A. \tag{42}$$

Proof: Taking g a constant equal to A, we have $S_n(g) = A$, and therefore we are looking for a sufficient condition guaranteeing that $S_n(f) - S_n(g)$ tends to 0 as n tends to ∞. By the localization principle, it suffices to show that

$$\lim_{n\uparrow\infty} \int_0^\delta \sin((n+\tfrac{1}{2})u)\,\frac{\phi(u)}{sin(u/2)}\,du = 0. \tag{43}$$

The two integrals in (41) and (43) differ by

$$\int_0^\delta \sin((n+\tfrac{1}{2})u)\,v(u)\,du, \tag{44}$$

where

$$v(u) = \phi(u)\left\{\frac{1}{u/2} - \frac{1}{\sin(u/2)}\right\}$$

is integrable on $[0, \delta]$. Therefore, by the Riemann–Lebesgue lemma, the quantity (44) tends to zero as $n \to \infty$. ■

Dini's Theorem

THEOREM **A3.4.** *Let f be a 2π-periodic locally integrable complex-valued function and let $x \in \mathbb{R}$. If for some $0 < \delta \le \pi$ and some $A \in \mathbb{R}$, the function*

$$t \to \frac{f(x+t) + f(x-t) - 2A}{t}$$

is integrable on $[0, \delta]$, *then*

$$\lim_{n\uparrow\infty} S_n^f(x) = A.$$

Proof: The hypothesis says that the function $\phi(u)/u$, where ϕ is defined in (42), is integrable, and therefore condition (41) of Theorem A3.3 is satisfied (Riemann–Lebesgue lemma). ■

We shall give two corollaries of Dini's result.

COROLLARY **A3.1.** *If a 2π-periodic locally integrable complex-valued function $f(t)$ is Lipschitz continuous of order $\alpha > 0$ about $x \in \mathbb{R}$, that is,*

$$|f(x+h) - f(x)| = O(|h|^\alpha) \quad \text{as } h \to 0,$$

then $\lim_{n\uparrow\infty} S_n^f(x) = f(x)$.

Proof: Indeed, with $A = f(x)$,

$$\left| \frac{f(x+t)+f(x-t)-2A}{t} \right| \leq K \frac{1}{|t|^{1-\alpha}},$$

for some constant K and for all t in a neighborhood of zero, and $1/|t|^{1-\alpha}$ is integrable in this neighborhood, because $1-\alpha < 1$. Dini's theorem A3.4 concludes the proof. ■

COROLLARY **A3.2.** *Let $f(t)$ be a 2π-periodic locally integrable complex-valued function, and let $x \in \mathbb{R}$ be such that*

$$f(x+0) = \lim_{h \downarrow 0} f(x+h) \quad \textit{and} \quad f(x-0) = \lim_{h \downarrow 0} f(x-h)$$

exist and are finite, and further assume that the derivatives to the left and to the right at x exist. Then

$$\lim_{n \uparrow \infty} S_n^f(x) = \frac{f(x+0)+f(x-0)}{2}.$$

Proof: By definition, one says that the derivative to the right exists if

$$\lim_{t \downarrow 0} \frac{f(x+t)-f(x+0)}{t}$$

exists and is finite, with a similar definition for the derivative to the left. The differentiability assumptions imply that

$$\lim_{t \downarrow 0} \frac{f(x+t)-f(x+0)+f(x-t)-f(x-0)}{t}$$

exists and is finite and therefore that

$$\frac{\phi(t)}{t} = \frac{f(x+t)+f(x-t)-2A}{t}$$

is integrable in a neighborhood of zero, where

$$2A = f(x+0)+f(x-0).$$

Dini's theorem A3.4 concludes the proof. ■

EXAMPLE **A3.1.** *Apply the previous theorem to the 2π-periodic function defined by*

$$f(t) = t \quad \textit{when } 0 < t \leq 2\pi.$$

One finds

$$t = \pi - \sum_{\substack{n \in \mathbb{Z} \\ n \neq 0}} 2 \frac{\sin(nt)}{n} \quad \textit{when } 0 < t < 2\pi.$$

For $t = 0$, we can directly check that the sum of the Fourier series is

$$\tfrac{1}{2}(f(0+)+f(0-)) = \tfrac{1}{2}(0+2\pi) = \pi,$$

as announced in the last corollary. For $t = \pi/2$, *we obtain the remarkable identity*

$$\frac{\pi}{4} = \frac{1}{1} - \frac{1}{3} + \frac{1}{5} - \frac{1}{7} + \cdots .$$

Jordan's Theorem

Jordan's convergence theorem features functions of bounded variation.

DEFINITION **A3.1.** *A real-valued function* $\varphi : \mathbb{R} \mapsto \mathbb{R}$ *is said to have* bounded variation *on the interval* $[a, b] \subset \mathbb{R}$ *if*

$$\sup_{\mathcal{D}} \sum_{i=0}^{n-1} |\varphi(x_{i+1}) - \varphi(x_i)| < \infty, \tag{45}$$

where the supremum is over all subdivisions $\mathcal{D} = \{a = x_0 < x_1 < \cdots < x_n = b\}$.

We quote without proof the fundamental result on the structure of bounded variation functions.

THEOREM **A3.5.** *A real-valued function* φ *has bounded variation over* $[a, b]$ *if and only if there exist two nondecreasing real-valued functions* φ_1, φ_2 *such that, for all* $t \in [a, b]$,

$$\varphi(t) = \varphi_1(t) - \varphi_2(t). \tag{46}$$

In particular, for all $x \in [a, b)$, φ has a limit to the right $\varphi(x + 0)$; for all $x \in (a, b]$, it has a limit to the left $\varphi(x - 0)$; and the discontinuity points of $\varphi(t)$ in $[a, b]$ form a denumerable set, and therefore a set of Lebesgue measure zero.

THEOREM **A3.6.** *Let* f *be a* 2π*-periodic locally integrable real-valued function of bounded variation in a neighborhood of a given* $x \in \mathbb{R}$. *Then*

$$\lim_{n\uparrow\infty} S_n^f(x) = \frac{f(x+0) + f(x-0)}{2}. \tag{47}$$

■

The proof is omitted.

EXERCISE **A3.2.** *Let* $f \in L^1_{\mathbb{C}}(\mathbb{R})$. *Show that, for any* $B > 0$,

$$\int_{-B}^{+B} \hat{f}(\nu) e^{2i\pi\nu t}\, d\nu = 2B \int_{\mathbb{R}} f(t+s) \operatorname{sinc}(2Bs)\, ds,$$

and use this to study the pointwise convergence of the left-hand side as B tends to infinity, along the lines of the current chapter.

The function

$$2B \operatorname{sinc}(2Bt)$$

is also called *Dirichlet's kernel*.

EXERCISE **A3.3.** *Let f_1 and f_2 be the 2π-periodic functions defined on $(-\pi, +\pi]$ by*

$$f_1(x) = x, \qquad f_2(x) = \pi^2 - 3x^2.$$

Compute their Fourier coefficients, and use this to compute

$$\sum_{n\geq 1} \frac{(-1)^n}{n}, \quad \sum_{n\geq 1} \frac{1}{n^2}, \quad \sum_{n\geq 1} \frac{1}{n^4}, \quad \sum_{n\geq 1} \frac{1}{n^6}.$$

Integration of Fourier Series

Let $f(t)$ be a real-valued 2π-periodic locally integrable function. Denoting by c_n the nth Fourier coefficient of $f(t)$, we have $c_{-n} = c_n^*$ because $f(t)$ is real. Therefore, the Fourier series of $f(t)$ can be written as

$$\tfrac{1}{2}a_0 + \sum_{n=1}^{\infty} \{a_n \cos(nx) + b_n \sin(nx)\}, \tag{48}$$

where, for $n \geq 1$,

$$a_n = \frac{1}{\pi}\int_0^{2\pi} f(t)\cos(nt)\,dt, \qquad b_n = \frac{1}{\pi}\int_0^{2\pi} f(t)\sin(nt)\,dt.$$

Of course, the series in (48) is purely formal when no additional constraints are put on $f(t)$ in order to guarantee its convergence. Now, the function $F(t)$ defined for $t \in [0, 2\pi)$ by

$$F(t) = \int_0^t (f(x) - \tfrac{1}{2}a_0)\,dx \tag{49}$$

is 2π-periodic, is continuous (observe that $F(0) = F(1) = 0$), and has bounded variation on finite intervals.

Therefore, by Jordan's theorem its Fourier series converges everywhere, and for all $x \in \mathbb{R}$,

$$F(x) = \tfrac{1}{2}A_0 + \sum_{n=1}^{\infty} \{A_n \cos(nx) + B_n \sin(nx)\},$$

where, for $n \geq 1$,

$$\begin{aligned} A_n &= \frac{1}{\pi}\int_0^{2\pi} F(t)\cos(nt)\,dt \\ &= \frac{1}{\pi}\left[F(x)\frac{\sin(nx)}{n}\right]_0^{2\pi} - \frac{1}{n\pi}\int_0^{2\pi} (f(t) - \tfrac{1}{2}a_0)\sin(nt)\,dt \\ &= -\frac{1}{n\pi}\int_0^{2\pi} f(t)\sin(nt)\,dt = -\frac{b_n}{n}, \end{aligned}$$

and, with a similar computation,

$$B_n = \frac{1}{\pi}\int_0^{2\pi} F(t)\sin(nt)\,dt = \frac{a_n}{n}.$$

Therefore, for all $x \in \mathbb{R}$,

$$F(x) = \tfrac{1}{2}A_0 + \sum_{n=1}^{\infty}\left\{\frac{a_n}{n}\sin(nx) - \frac{b_n}{n}\cos(nx)\right\}. \tag{50}$$

The constant A_0 is identified by setting $x = 0$ in (50):

$$\tfrac{1}{2}A_0 = \sum_{n=1}^{\infty}\frac{b_n}{n}. \tag{51}$$

Since A_0 is finite we have shown, in particular, that $\sum_{n=1}^{\infty} b_n/n$ converges for any sequence $\{b_n\}_{n\geq 1}$ of the form

$$b_n = \frac{1}{\pi}\int_0^{2\pi} f(t)\sin(nt)\,dt,$$

where, $f(t)$ is a real function integrable over $[0, 2\pi]$.

Gibbs' Overshoot Phenomenon

We close this section by mentioning a phenomenon typical of the behavior of a Fourier series at a discontinuity of the function. *Gibbs' overshoot phenomenon* has nothing to do with the failure of the Fourier series to converge at a point of discontinuity of the corresponding function. It concerns the overshoot of the partial sums at such a point of discontinuity. An example will demonstrate this effect.

Consider the 2π-periodic function defined in the interval $(-\pi, +\pi]$ by

$$f(x) = \begin{cases} \dfrac{\pi}{2} - \dfrac{x}{2} & \text{if } x > 0, \\ 0 & \text{if } x < 0, \\ -\dfrac{\pi}{2} - \dfrac{x}{2} & \text{if } x < 0. \end{cases}$$

The partial sum of its Fourier series is

$$S_n^f(x) = \sum_{k=1}^{n}\frac{\sin(nx)}{n}.$$

By Dini's theorem, the partial sum $S_n^f(0)$ converges pointwise to $(1/2)(f(0+) + f(0-)) = \pi/2$. However, we shall see that for some $A > \pi/2$ and sufficiently large n,

$$S_n^f\left(\frac{\pi}{n}\right) \geq A. \tag{52}$$

Therefore, there exist a constant $c > 0$ and a neighborhood N_0 of 0 such that $|S_n^f(x) - S_n^f(0)| \geq c$ whenever $x \in N_0 - \{0\}$. This constitutes Gibbs' overshoot phenomenon, which can be observed whenever the function has a point of discontinuity. The proof of (52) for this special case keeps most of the features of the general proof, which is left for the reader. In this special case,

$$S_n^f(x) = \int_0^x \frac{\sin((n+\frac{1}{2})t)}{2\sin(\frac{1}{2}t)}\,dt - \frac{x}{2}.$$

Now,

$$\begin{aligned}
&\int_0^x \frac{\sin((n+\frac{1}{2})t)}{2\sin(\frac{1}{2}t)}\,dt \\
&= \int_0^x \left(\frac{\sin(nt)\cos(\frac{1}{2}t)}{2\sin(\frac{1}{2}t)} + \frac{1}{2}\cos(nt)\right)dt \\
&= \int_0^x \frac{\sin(nt)}{t}\,dt + \int_0^x \sin(nt)\left(\frac{\cos(\frac{1}{2}t)}{2\sin(\frac{1}{2}t)} - \frac{1}{t}\right)dt \\
&\quad + \frac{1}{2}\int_0^x \cos(nt)\,dt.
\end{aligned}$$

The last two integrals converge uniformly to zero (by the uniform version of the Riemann–Lebesgue lemma). Also,

$$\int_0^{\frac{\pi}{n}} \frac{\sin(nt)}{t}\,dt = \int_0^{\pi} \frac{\sin(t)}{t}\,dt \simeq 1.18\,\frac{\pi}{2} > \frac{\pi}{2}.$$

■

A3·2 Féjer's Theorem

The Fourier $S_n^f(t)$ series of a 2π-periodic locally integrable function f converges to $f(t)$ for a given t only under certain conditions (see the previous section). However, Cesaro convergence of the series requires much milder conditions. For a 2π-periodic locally stable function f, Féjer's sum

$$\sigma_n^f(t) = \frac{1}{n}\sum_{k=0}^{n-1} S_k^f(t) \tag{53}$$

behaves more nicely than the Fourier series itself. In particular, for continuous functions, it converges pointwise to the function itself. Therefore, Féjer's theorem is a kind of inversion formula, in that it shows that for a large class of periodic functions (see the precise statement in Theorem A3.11 ahead), the function can be recovered from its Fourier coefficients.

Féjer's Kernel

Take the imaginary part of the identity

$$\sum_{k=0}^{n-1} e^{i(k+1/2)u} = e^{iu/2}\frac{1-e^{inu}}{1-e^{iu}}$$

to obtain

$$\sum_{k=0}^{n-1} \sin((k+\tfrac{1}{2})u) = \frac{\sin^2(\frac{1}{2}nu)}{\sin(\frac{1}{2}u)}.$$

Starting from Dirichlet's integral expression for $S_n^f(t)$ [*cf.*, Eq. (37)], we obtain, in view of the identity just proven, Féjer's integral representation of $\sigma_n^f(x)$,

$$\sigma_n^f(x) = \int_{-\pi}^{+\pi} K_n(u)\, f(x-u)\, du = \int_{-\pi}^{+\pi} K_n(x-u)\, f(x)\, du, \tag{54}$$

where

$$K_n(t) = \frac{1}{2n\pi}\frac{\sin^2(\frac{1}{2}nt)}{\sin^2(\frac{1}{2}t)} \tag{55}$$

is, by definition, *Féjer's kernel*. It has the following properties:

$$K_n(t) \geq 0, \tag{56}$$

and [letting $f(t) = 1$ in (54)],

$$\int_{-\pi}^{+\pi} K_n(u)\, du = 1. \tag{57}$$

Also (the proofs are left as an exercise),

$$\lim_{n\uparrow\infty} K_n(t) = 1, \tag{58}$$

and, for all $\varepsilon \leq \pi$,

$$\lim_{n\uparrow\infty} \int_{-\varepsilon}^{+\varepsilon} K_n(u)\, du = 1. \tag{59}$$

The last four properties make of Féjer's kernel a regularization kernel on $[-\pi, +\pi]$ (by definition of a regularization kernel).

Cesaro Convergence for Fourier Series of Continuous Functions

We first treat the case of continuous functions, because the result can be obtained from the basic principles of analysis, in particular, without recourse to the Riemann–Lebesgue lemma.

THEOREM **A3.7.** *Let $f(t)$ be a 2π-periodic continuous function. Then*

$$\lim_{n\uparrow\infty} \sup_{x\in[-\pi,+\pi]} |\sigma_n^f(x) - f(x)| = 0. \tag{60}$$

Proof: From (54) and (56), we have

$$|\sigma_n^f(x) - f(x)| \le \int_{-\pi}^{+\pi} K_n(u)\, |f(x-u) - f(x)|\, du$$

$$= \int_{-\delta}^{+\delta} + \int_{[-\pi,+\pi]\setminus[-\delta,+\delta]} = A + B. \tag{61}$$

For a given $\varepsilon > 0$, choose δ such that $|f(x-u) - f(x)| \le \varepsilon/2$ when $|u| \le \delta$. Note that f is uniformly continuous and uniformly bounded (being a periodic and continuous function), and therefore δ can be chosen independently of x. We have

$$A \le \frac{\varepsilon}{2} \int_{-\delta}^{+\delta} K_n(u)\, du \le \frac{\varepsilon}{2},$$

and, calling M the uniform bound of f,

$$B \le 2M \int_{[-\pi,+\pi]\setminus[-\delta,+\delta]} K_n(u)\, du.$$

By (57) and (59), $B \le \varepsilon/2$ for n sufficiently large. Therefore, for n sufficiently large, $A + B \le \varepsilon$. ■

Féjer's theorem for continuous periodic functions is the key to important approximation theorems. The first one is for free. We call a trigonometric polynomial any finite trigonometric sum of the form

$$p(x) = \sum_{-n}^{+n} c_k e^{ikx}.$$

THEOREM **A3.8.** *Let $f(t)$ be a 2π-periodic continuous function. Select an arbitrary $\varepsilon > 0$. Then there exists a trigonometric polynomial $p(x)$ such that*

$$\sup_{t\in[-\pi,+\pi]} |f(x) - p(x)| \le \varepsilon.$$

Proof: Use Theorem A3.7 and observe that $\sigma_n^f(x)$ is a trigonometric polynomial. ■

From this, we obtain the *Weierstrass approximation theorem.*

THEOREM **A3.9.** *Let $f : [a,b] \mapsto \mathbb{C}$ be a continuous function. Select an arbitrary $\varepsilon > 0$. There exists a polynomial $P(x)$ such that*

$$\sup_{t\in[a,b]} |f(x) - P(x)| \le \varepsilon.$$

If, moreover, f is real-valued, then P can be chosen with real coefficients.

Proof: First, suppose that $a = 0$, $b = 1$. One can then extend $f : [0,1]] \mapsto \mathbb{C}$ to a function still denoted by f, $f : [-\pi,+\pi]] \mapsto \mathbb{C}$, that is continuous and such that $f(+\pi) = f(-\pi) = 0$. By Theorem A3.8, there exists a trigonometric polynomial $p(x)$ such that

$$\sup_{t\in[0,1]} |f(x)-p(x)| \leq \sup_{t\in[-\pi,+\pi]} |f(x)-p(x)| \leq \frac{\varepsilon}{2}.$$

Now replace each term e^{ikx} in $p(x)$ by a sufficiently large portion of its Taylor series expansion, to obtain a polynomial $P(x)$ such that

$$\sup_{t\in[0,1]} |P(x)-p(x)| \leq \frac{\varepsilon}{2}.$$

Then

$$\sup_{t\in[0,1]} |f(x)-P(x)| \leq \frac{\varepsilon}{2}.$$

To treat the general case $f : [a,b] \mapsto \mathbb{C}$, apply the result just proven to $\varphi : [0,1] \mapsto \mathbb{C}$ defined by $\varphi(t) = f(a+(b-a)t)$ to obtain the approximating polynomial $\pi(x)$, and take $P(x) = \pi((x-a)/(b-a))$.

Finally, to prove the last statement of the theorem, observe that

$$|f(x) - \text{Re } P(x)| \leq |f(x)-P(x)|. \qquad \blacksquare$$

Féjer's Theorem

We shall first obtain for the Féjer's sum the result analogous to Theorem A3.3. First, from (54), we obtain

$$\sigma_n^f(x) = \frac{1}{2n\pi}\int_0^\pi \frac{\sin^2(\frac{1}{2}nu)}{\sin^2(\frac{1}{2}u)}\{f(x+u)-f(x-u)\}\,du; \tag{62}$$

therefore, for any number A,

$$\sigma_n^f(x) - A = \frac{1}{2n\pi}\int_0^\pi \frac{\sin^2(\frac{1}{2}nu)}{\sin^2(\frac{1}{2}u)}\{f(x+u)+f(x-u)-2A\}\,du. \tag{63}$$

THEOREM **A3.10.** *For any $x \in \mathbb{R}$ and any constant A,*

$$\lim_{n\uparrow\infty} \sigma_n^f(x) = A \tag{64}$$

if, for some $\delta > 0$,

$$\lim_{n\uparrow\infty} \frac{1}{n}\int_0^\delta \sin^2(\tfrac{1}{2}nu)\,\frac{\phi(u)}{u^2}\,du = 0, \tag{65}$$

where

$$\phi(u) = f(x+u)+f(x-u)-2A. \tag{66}$$

Proof: The quantity

$$\left|\frac{1}{n}\int_\delta^\pi \frac{\sin^2(\frac{1}{2}nu)}{\sin^2(\frac{1}{2}u)}\phi(u)\,du\right| \leq \frac{1}{n}\int_\delta^\pi \frac{|\phi(u)|}{\sin^2(\frac{1}{2}u)}\,du$$

tends to 0 as $n \uparrow \infty$. We must therefore show that

$$\frac{1}{n}\int_0^{\delta} \frac{\sin^2(\frac{1}{2}nu)}{\sin^2(\frac{1}{2}u)}\,\phi(u)\,du$$

tends to 0 as $n \uparrow \infty$. However, (65) guarantees this because

$$\left|\frac{1}{n}\int_0^{\delta} \sin^2(\tfrac{1}{2}nu)\left\{\frac{1}{\sin^2(\frac{1}{2}u)} - \frac{1}{\frac{1}{2}u^2}\right\}\phi(u)\,du\right|$$

$$\leq \frac{1}{n}\int_0^{\delta}\left\{\frac{1}{\sin^2(\frac{1}{2}u)} - \frac{1}{\frac{1}{2}u^2}\right\}|\phi(u)|\,du$$

tends to 0 as $n \uparrow \infty$ (the expression in curly brackets is bounded in $[0, \delta]$, and therefore the integral is finite). ■

THEOREM **A3.11.** *Let $f(t)$ be a 2π-periodic locally integrable function and assume that, for some $x \in \mathbb{R}$, the limits to the right and to the left (respectively, $f(x+0)$ and $f(x-0)$), exist. Then*

$$\lim \sigma_n^f(x) = \frac{f(x+0) + f(x-0)}{2}. \tag{67}$$

Proof: Fix $\delta > 0$. In view of the last result, it suffices to prove (65) with

$$\phi(u) = \{f(x+u) - f(x+0)\} + \{f(x-u) - f(x-0)\}.$$

Since $\phi(u)$ tends to 0 as $n \to \infty$, for any given $\varepsilon > 0$ there exists $\eta = \eta(\varepsilon)$, $0 < \eta \leq \delta$, such that $|\phi(u)| \leq \varepsilon$ when $0 < u \leq \eta$. Now,

$$\left|\frac{1}{n}\int_0^{\delta} \frac{\sin^2(\frac{1}{2}nu)}{u^2}\,\phi(u)\,du\right|$$

$$\leq \frac{1}{n}\int_0^{\eta} \frac{\sin^2(\frac{1}{2}nu)}{u^2}\,\varepsilon\,du + \frac{1}{n}\int_{\eta}^{\delta} \frac{\sin^2(\frac{1}{2}nu)}{u^2}\,|\phi(u)|\,du$$

$$\leq \frac{\varepsilon}{n}\int_0^{\eta} \frac{\sin^2(\frac{1}{2}nu)}{u^2}\,du + \frac{1}{n}\int_{\eta}^{\delta} \frac{|\phi(u)|}{u^2}\,du.$$

The last integral is bounded; and therefore, the last term goes to 0 as $n \uparrow \infty$. As for the penultimate term, it is bounded by $A\varepsilon$, where

$$A = \int_0^{\infty} \frac{\sin^2(\frac{1}{2}v)}{v^2}\,dv < \infty.$$ ■

A3·3 The Poisson Formula

The following corollary of Féjer's theorem will play the key role for the proof of the Poisson sum formula (Theorem A3.12).

COROLLARY **A3.1.** *Let f be a 2π-periodic locally integrable function and suppose that, for some $x \in \mathbb{R}$,*

(a) the function f is continuous at x, and

(b) its Fourier series $S_n^f(x)$ converges to some number A.

Then $A = f(x)$.

Proof: From (b) we see that

$$\lim_{n\uparrow\infty} \sigma_n^f(x) = A.$$

From Féjer's theorem and (a),

$$\lim_{n\uparrow\infty} \sigma_n^f(x) = f(x).$$ ■

We have already given a weak version of the Poisson sum formula in Section A2·2. A most interesting situation is when the function $\Phi(t)$ defined by (31) is equal to its Fourier series for *all* $t \in \mathbb{R}$, that is,

$$\sum_{n\in\mathbb{Z}} s(t+nT) = \frac{1}{T}\sum_{n\in\mathbb{Z}} \hat{s}\left(\frac{n}{T}\right) e^{2i\pi\frac{n}{T}t} \quad \text{for all } t \in \mathbb{R}. \tag{68}$$

The next theorem extends the result in Exercise A2.8.

THEOREM **A3.12.** *Let $s(t)$ be a stable complex signal, and let* $0 < T < \infty$ *be fixed. Assume in addition that*

(1) $\sum_{n\in\mathbb{Z}} s(t+nT)$ converges everywhere *to some* continuous *function,*

(2) $\sum_{n\in\mathbb{Z}} \hat{s}\left(\frac{n}{T}\right) e^{2i\pi\frac{n}{T}t}$ converges for all t.

Then the strong Poisson formula (68) *holds.*

Proof: The result is an immediate consequence of both the weak Poisson summation result (Theorem A2.3) and the corollary of Féjer's theorem in Section A3·2. ■

Here are two important cases for which the strong Poisson sum formula holds.

COROLLARY **A3.1.** *Let $s(t)$ be a stable complex signal, and let* $0 < T < \infty$ *be fixed. If, in addition, $\sum s(t+nT)$ converges* everywhere *to a* continuous *function that has* bounded variation, *then the Poisson formula* (68) *holds.*

Proof: We must verify conditions (1) and (2) of Theorem A3.12. Condition (1) is part of the hypothesis. Condition (2) is a consequence of Jordan's theorem A3.6. ■

EXAMPLE **A3.1.** *If $s(t)$ is continuous, has bounded support, and has bounded variation, the Poisson sum formula* (68) *holds.*

COROLLARY **A3.2.** *If a stable continuous signal $s(t)$ satisfies*

$$s(t) = O\left(\frac{1}{1+|t|^\alpha}\right) \quad \text{as } |t| \to \infty,$$
$$\hat{s}(\nu) = O\left(\frac{1}{1+|\nu|^\alpha}\right) \quad \text{as } |\nu| \to \infty, \tag{69}$$

for some $\alpha > 1$, then the Poisson formula (68) *holds for all* $0 < T < \infty$.

Proof: The result is an immediate corollary of Theorem A3.12. ■

Convergence Improvement

The Poisson formula can be used to replace a series with slow convergence by one with rapid convergence, or to obtain some remarkable formulas. Here is a typical example. For $a > 0$,

$$s(t) = e^{-2\pi a|t|} \overset{\text{FT}}{\rightarrow} \hat{s}(\nu) = \frac{a}{\pi(a^2+\nu^2)}.$$

Since

$$\sum_{n\in\mathbb{Z}} s(t+n) = \sum_{n\in\mathbb{Z}} e^{-2\pi a|t+n|}$$

is a continuous function with bounded variation, we have the Poisson formula, that is,

$$\sum_{n\in\mathbb{Z}} e^{-2\pi a|n|} = \sum_{n\in\mathbb{Z}} \frac{a}{\pi(a^2+n^2)}.$$

The left-hand side is equal to

$$\frac{2}{1-e^{-2\pi a}} - 1,$$

and the right-hand side can be written as

$$\frac{1}{\pi a} + 2\sum_{n\geq 1} \frac{a}{\pi(a^2+n^2)}.$$

Therefore,

$$\sum_{n\geq 1} \frac{1}{a^2+n^2} = \frac{\pi}{2a}\frac{1+e^{-2\pi a}}{1-e^{-2\pi a}} - \frac{1}{2a^2}.$$

Letting $a \to 0$, we have

$$\sum_{n\geq 1} \frac{1}{n^2} = \frac{\pi^2}{6}.$$

The general feature of the above example is the following. We have a series that is obtained by sampling a very regular function (in fact, C^∞) but also slowly

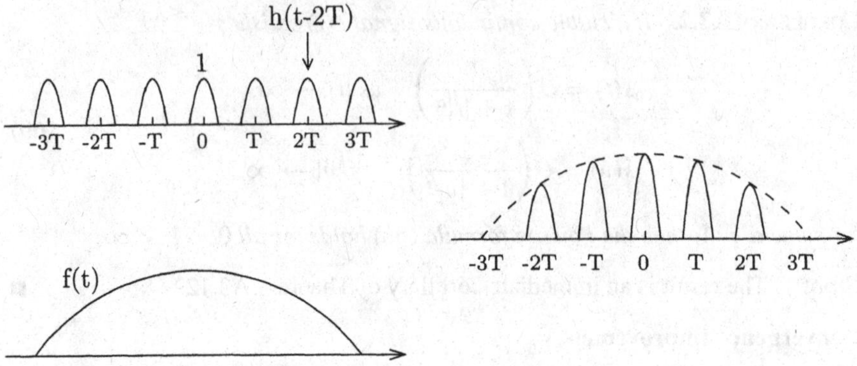

Figure A3.1. Radar return signal

decreasing. However, because of its strong regularity, its FT has a fast decay. The series obtained by sampling the FT is therefore quickly converging.

Radar Return Signal

Let $s(t)$ be a signal of the form

$$s(t) = \left(\sum_{n\in\mathbb{Z}} h(t - nT)\right) f(t). \tag{70}$$

(We may interpret $h(t - nT)$ as a return signal of the nth pulse of a radar after reflection on the target, and $f(t)$ as a modulation due to the rotation of the antenna.) The FT of this signal is (see Fig. A3.1)

$$\hat{s}(\nu) = \frac{1}{T}\sum_{n\in\mathbb{Z}} \hat{h}\left(\frac{n}{T}\right) \hat{f}\left(\nu - \frac{n}{T}\right). \tag{71}$$

EXERCISE **A3.4.** *Show that if (1) $f(t)$ is integrable, (2) $\sum_{n\in\mathbb{Z}} h(t-nT)$ is integrable and continuous, and (3) $\sum_{n\in\mathbb{Z}} \hat{h}(n/T) < \infty$, then* (71) *holds true. Find other conditions.*

References

[A1] Ablowitz, M.J. and Jokas, A.S. (1997). *Complex Variables*, Cambridge University Press.

[A2] Bracewell, R.N. (1991). *The Fourier Transform and Its Applications*, 2nd rev. ed., McGraw-Hill; New York.

[A3] Gasquet, C. and Witomski, P. (1991). *Analyse de Fourier et Applications*, Masson: Paris.

[A4] Helson, H. (1983). *Harmonic Analysis*, Addison-Wesley: Reading, MA.

[A5] Katznelson, Y. (1976). *An Introduction to Harmonic Analysis*, Dover: New York.

[A6] Kodaira, K. (1984). *Introduction to Complex Analysis*, Cambridge University Press.

[A7] Körner, T.W. (1988). *Fourier Analysis*, Cambridge University Press.

[A8] Rudin, W. (1966). *Real and Complex Analysis*, McGraw-Hill: New York.
[A9] Titchmarsh, E.C. (1986). *The Theory of Functions*, Oxford University Press.
[A10] Tolstov, G. (1962). *Fourier Series*, Prentice-Hall (Dover edition, 1976).
[A11] Zygmund, A. (1959). *Trigonometric Series*, (2nd ed., Cambridge University Press.

Part B

Signal Processing

Introduction

The Fourier transform derives its importance in physics and in electrical engineering from the fact that many devices mapping an input signal $x(t)$ into an output signal $y(t)$ have the following property: If the input is a complex sinusoid $e^{2i\pi\nu t}$, the output is $T(\nu)e^{2i\pi\nu t}$, where $T(\nu)$ is a complex function characterizing the device. For example, when $x(t)$ and $y(t)$ are, respectively, the voltage observed at the input and the steady-state voltage observed at the output of an RC circuit (see Fig. B0.1), the input–output mapping takes the form of a linear differential equation:

$$y(t) + RC\dot{y}(t) = x(t),$$

and it can be readily checked that

$$T(\nu) = \frac{1}{1 + 2i\pi\nu RC}.$$

The RC circuit is one of the physical devices that transform a signal into another signal, that satisfy the superposition principle, and that are time-invariant. More precisely:

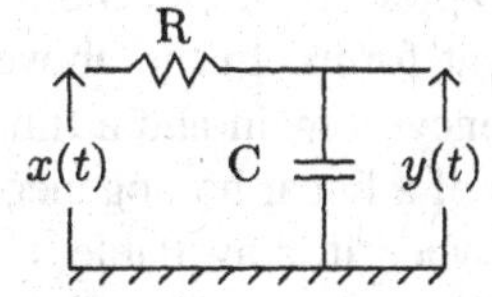

Figure B0.1. The RC circuit

(a) If $y_1(t)$ and $y_2(t)$ are the outputs corresponding to the inputs $x_1(t)$ and $x_2(t)$, then $\lambda_1 y_1(t) + \lambda_2 y_2(t)$ is the output corresponding to the input $\lambda_1 x_1(t) + \lambda_2 x_2(t)$;

(b) If $y(t)$ is the output corresponding to $x(t)$, then $y(t - \tau)$ is the output corresponding to $x(t - \tau)$.

Such physical devices are called (homogeneous linear) filters.

A basic example is the convolutional filter, for which the input–output mapping takes, in the time domain, the form

$$y(t) = \int_{\mathbb{R}} h(t - s)x(s)\,\mathrm{d}s,$$

where $h(t)$ is called the impulse response, because it is the response of the filter when the Dirac pulse $\delta(t)$ is applied at the input. Indeed,

$$h(t) = \int_{\mathbb{R}} h(t - s)\delta(s)\,\mathrm{d}s.$$

If the impulse response is integrable, the output is well defined and integrable, as long as the input is integrable. Then, by the convolution–multiplication rule, the expression of the input–output mapping in the frequency domain is

$$\hat{y}(\nu) = T(\nu)\hat{x}(\nu),$$

where $T(\nu)$ is the frequency response, that is, the FT of the impulse response:

$$T(\nu) = \int_{\mathbb{R}} h(t)\, e^{-2i\pi\nu t}\,\mathrm{d}t.$$

Observe that if the input is $x(t) = e^{-2i\pi\nu t}$, the output is well defined and equal to

$$\int_{\mathbb{R}} h(s)x(t - s)\,\mathrm{d}s = \int_{\mathbb{R}} h(s)\, e^{-2i\pi\nu(t-s)}\mathrm{d}s = T(\nu)e^{-2i\pi\nu t},$$

in accordance with what was said in the beginning of this introduction.

In the particular case of the RC filter, the solution of the differential equation with arbitrary initial condition at $-\infty$ is indeed of the convolution type, with the impulse response

$$h(t) = e^{-\frac{t}{RC}}\, 1_{\mathbb{R}_+}(t).$$

The RC filter is a convolutional filter, and it contains the typical features of the more general filters. In the general case, since a filter is a mapping, we shall have to define its domain of application. Depending on this domain, the input–output mapping takes different forms. In the above informal discussion of the RC circuit, there are a frequency-domain and a time-domain representation and also a representation in terms of a linear homogeneous differential equation. The latter is not general. In fact, when it is available, the transmittance is a rational function of the frequency ν. The corresponding filters, called *rational filters*, form an important class, and Chapter B1 gives the basic concepts concerning analog (that is, continuous-time) filters.

In addition to filtering, there are two fundamental operations of interest in communications systems: frequency transposition and sampling. Frequency transposition is a basic technique of analog communications. It has two main applications, the first of which is transmission. Indeed, the Hertzian channels are in the high-frequency bands—in fact, much higher than the one of brute signals such as the electric signals carrying voice, for instance—and consequently, the latter have to be frequency-shifted. The second reason is resource utilization and is related to frequency multiplexing, a technique by which signals initially occupying the same frequency band are shifted to nonoverlapping bands and can then be simultaneously transmitted without mutual interference. From a mathematical point of view the theory of frequency transposition (or, equivalently, of band-pass signals, to be defined in Chapter B1) is not difficult. It remains interesting because of the special phenomena associated with this technique, such as cross-talk in quadrature multiplexing and dispersion phenomena.

In digital communications systems, an analog signal $s(t)$ is first sampled, and the result is a sequence of samples $s(nT)$, $n \in \mathbb{Z}$. It is important to identify conditions under which the sample sequence faithfully represents the original signal. The central result of Chapter B2 is the so-called Shannon–Nyquist theorem, which says that this is true if the signal $s(t)$ is stable and continuous and if the support of its FT $\hat{s}(\nu)$ is contained in the interval $[-1/T, +1/T]$. The original signal can then be recovered by the reconstruction formula:

$$s(t) = \sum_{n\in\mathbb{Z}} s(nT)\,\mathrm{sinc}\left(\frac{t}{T} - n\right).$$

The theory of sampling is an application of the results obtained in Part A, and in particular of the Poisson sum formula. The above reconstruction formula has many sources,[1] and its importance in communications was fully realized by Claude Shannon and Harald Nyquist.

The Shannon–Nyquist sampling theorem is the bridge between the analog (physical) world and the discrete-time (computational) world of digital signal processing. The reader will find in the main text a brief discussion of the interest of digital communications systems. Therefore, a large portion of this Part B is devoted to discrete-time signals (Chapters B2–B4).

As we have already mentioned, the Poisson sum formula is the key to the sampling theorem. It also plays a very important role in the numerical analysis of the discrete Fourier transform considered as an approximation of the continuous Fourier transform (see Chapter B3) and also in the intersymbol interference problem (see Chapter B2).

The study of the interaction between discrete time and continuous time is not limited to the sampling theorem. For instance, we prove that filtering and sampling

[1]See J.R. Higgins, Five short stories about the cardinal series, *Bull. Amer. Math. Soc.*, 12, 1985, 45–89.

commute for base-band signals. This is not a difficult result, but it is of course a fundamental one because in signal processing, one first samples and then performs the filtering operation in the sampled domain, since one of the advantages of digital processing comes precisely from the difficulty of making analog filters.

One advantage of analog processing is that it is instantaneous. To maintain competitivity, the signal processing algorithms have to be fast. For instance, the discrete Fourier transform is implemented by the so-called fast Fourier transform, an algorithm whose principle we briefly explain in this Part. Subband coding also has a fast algorithm associated with it. It is a data compression technique. The signal is not directly quantized, but instead, it is first analyzed by a filter bank, and the output of each filter bank is quantized separately. This allows one to dispatch the compression resources unequally, with fewer bits allocated to the subbands that are less informative (see the discussion in Chapter B4). Subband coding is the last topic of Part B and introduces the sections on multiresolution analysis in Part D.

B1

Filtering

B1·1 Impulse Response and Frequency Response

Convolutional Filter

We introduce a particular and very important class of filters.

DEFINITION **B1.1.** *The transformation from the stable signal $x(t)$ to the stable signal $y(t)$ defined by the convolution*

$$y(t) = \int_{\mathbb{R}} h(t-s)x(s)\,ds, \tag{1}$$

where $h(t)$ is stable, is called a convolutional filter . *This filter is called a* causal filter *if $h(t) = 0$ for $t < 0$.*

The signal $y(t)$ is the *output*, whereas the signal $x(t)$ is the *input* of the linear filter with *impulse response* $h(t)$. Informally, if $x(t)$ is the Dirac generalized function $\delta(t)$ (an impulse at time 0), the output is (see Fig. B1.1)

$$\int_{\mathbb{R}} h(t-s)\delta(s)\,ds = h(t),$$

whence the terminology.

A causal filter responds only after it has been stimulated. For this reason, it is sometimes also called a *realizable filter* (Fig. B1.1. features a causal impulse response). For such filters, the input–output relationship (1) becomes (note the

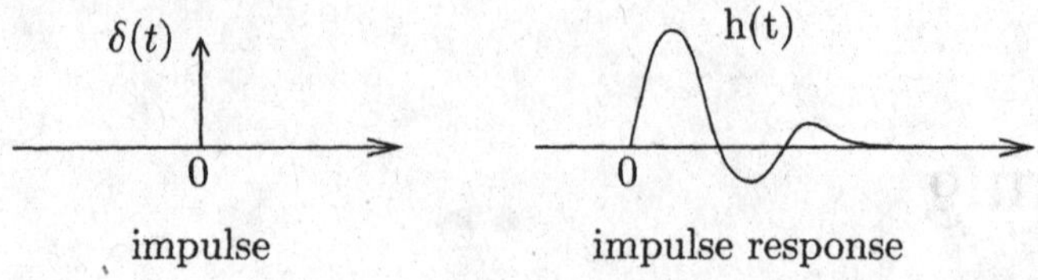

Figure B1.1. Impulse and impulse response

upper limit of integration)

$$y(t) = \int_{-\infty}^{t} h(t-s)x(s)\,ds. \tag{2}$$

DEFINITION **B1.2.** *The Fourier transform of the (stable) impulse response $h(t)$,*

$$T(\nu) = \int_{\mathbb{R}} h(t)e^{-2i\pi\nu t}\,dt, \tag{3}$$

is called the frequency response.

If the input is the complex sinusoid $x(t) = e^{2i\pi\nu t}$, by (1), the output is

$$y(t) = T(\nu)e^{2i\pi\nu t}. \tag{4}$$

(Note that the output is well defined by the convolution formula, even though in this particular case the input is not integrable.)

EXERCISE **B1.1.** *Let $y(t)$ be the output of a stable and causal convolutional filter with impulse response $h(t)$ [see (2)]. Let*

$$z(t) = \int_{0}^{t} h(t-s)x(s)\,ds, \qquad t \geq 0,$$

be the output of the same filter, when the input $x(t)$ is applied only from time $t = 0$ on. Show that

$$\lim_{t\uparrow+\infty} |z(t) - y(t)| = 0.$$

A More General Definition

Convolutional filters are only a special class of filters. A more general definition is as follows. Denote by $\mathbb{C}^{\mathbb{R}}$ the set of functions of $\mathbb{R}$ into $\mathbb{C}$.

DEFINITION **B1.3.** *Let $D(\mathcal{L})$ be a set of functions from $\mathbb{R}$ into $\mathbb{C}$ with the two following properties:*

(α) It is closed under linear operations;
(β) it is closed under translation.

$\mathcal{L} : D(\mathcal{L}) \mapsto \mathbb{C}^{\mathbb{R}}$ *is called a* homogeneous linear filter *with* domain $D(\mathcal{L})$ *if:*

(i) $\mathcal{L}$ is linear, and
(ii) $\mathcal{L}$ is time-invariant.

The meaning of properties (α) and (β) is the following: (α) $x_1(t), x_2(t) \in D(\mathcal{L})$, $\lambda_1, \lambda_2 \in \mathbb{C} \implies \lambda_1 x_1(t) + \lambda_2 x_2(t) \in D(\mathcal{L})$; and ($\beta$) $x(t) \in D(\mathcal{L}), T \in \mathbb{R} \implies x(t-T) \in D(\mathcal{L})$.

The meaning of properties (i) and (ii) is the following: (i) $x_1(t), x_2(t) \in D(\mathcal{L})$, $\lambda_1, \lambda_2 \in \mathbb{C}, x_1(t) \overset{\mathcal{L}}{\to} y_1(t), x_2(t) \overset{\mathcal{L}}{\to} y_2(t) \implies \lambda_1 x_1(t) + \lambda_2 x_2(t) \overset{\mathcal{L}}{\to} \lambda_1(t) y_1(t) + \lambda_2 y_2(t)$; (ii) $x(t) \in D(\mathcal{L}), T \in \mathbb{R}, x(t) \overset{\mathcal{L}}{\to} y(t) \implies x(t-T) \overset{\mathcal{L}}{\to} y(t-T)$.

EXERCISE **B1.2.** *Show that if $e^{2i\pi\nu t} \in D(\mathcal{L})$, then*

$$e^{2i\pi\nu t} \overset{\mathcal{L}}{\to} T(\nu) e^{2i\pi\nu t} \tag{5}$$

for some complex number $T(\nu)$.

The function $T(\nu)$ is called the *frequency response* of the filter. Every frequency response is of the form

$$T(\nu) = G(\nu) e^{i\beta(\nu)}, \tag{6}$$

where $G(\nu) = |T(\nu)|$ is the *amplitude gain* and $\beta(\nu) = \operatorname{Arg} T(\nu)$ is the *phase.*

EXAMPLE **B1.1.** *Let*

$$D(\mathcal{L}) = \{x(t) : \int_{\mathbb{R}} |x(t)|\, dt < \infty\},$$

or

$$D(\mathcal{L}) = \{x(t) + e(t) : \int_{\mathbb{R}} |x(t)|\, dt < \infty \text{ and } e(t) \in \mathcal{E}\},$$

where $\mathcal{E}$ is the set of complex finite linear combinations of complex exponentials. For any signal in $D(\mathcal{L})$, the right-hand side of (1) *is well defined, and we can therefore define the filter $\mathcal{L}$ with domain $D(\mathcal{L})$ by the input–output relationship* (1). *The frequency response, as defined by* (5), *is then the FT of $h(t)$.*

EXAMPLE **B1.2.** *Let $h(t) \in L^2_{\mathbb{C}}(\mathbb{R})$. We shall see in Part C that the FT $\hat{h}(\nu) = T(\nu)$ of $h(t)$ can be defined and that it is in $L^2_{\mathbb{C}}(\mathbb{R})$. We take $D(\mathcal{L}) = L^2_{\mathbb{C}}(\mathbb{R})$ and define $\mathcal{L}$ by the input–output relationship*

$$y(t) = \int_{\mathbb{R}} T(\nu) \hat{x}(\nu) e^{2i\pi\nu t}\, d\nu, \tag{7}$$

where $\hat{x}(\nu)$ is the FT of the input $x(t) \in L^2_{\mathbb{C}}(\mathbb{R})$. The right-hand side of (7) *has a meaning since $T(\nu)$ and $\hat{x}(\nu)$ being in $L^2_{\mathbb{C}}(\mathbb{R})$ implies that $T(\nu)\hat{x}(\nu)$ is in $L^1_{\mathbb{C}}(\mathbb{R})$ (see Theorem 20 of the appendix).*

EXAMPLE **B1.3.** *If $T(\nu)$ is an arbitrary function, not necessarily in $L^2_{\mathbb{C}}(\mathbb{R})$, one can always define a filter $\mathcal{L}$ by the input–output relation* (7), *provided one chooses for domain $D(\mathcal{L})$ the set of signals $x(t)$ such that the right-hand side has a meaning.*

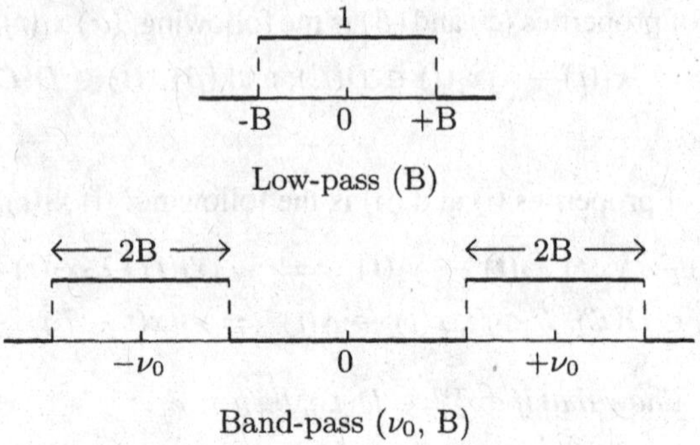

Figure B1.2. Low-pass and band-pass frequency responses

Low-Pass, Band-Pass, and Hilbert Filters

The low-pass and band-pass filters (see Fig. B1.2) that we now define belong to the category of Example B1.2.

One calls *low-pass* (B) a filter with frequency response

$$T(\nu) = 1_{[-B,+B]}(\nu), \tag{8}$$

where B is the *cut-off frequency*. One calls *band-pass* (B, ν_0), where $0 < B < \nu_0$, a filter with frequency response

$$T(\nu) = 1_{[-\nu_0-B,-\nu_0+B]}(\nu) + 1_{[\nu_0-B,\nu_0+B]}(\nu), \tag{9}$$

where ν_0 is the *center frequency*, and $2B$ is the *bandwidth*.

Hilbert's filter (see Fig. B1.3) belongs to the category of Example B1.3. It is the filter with frequency response

$$T(\nu) = i \operatorname{sgn}(\nu), \quad \text{where } T(0) = 0. \tag{10}$$

One possible domain for Hilbert's filter is the set of stable (resp., finite-energy) signals whose FT has compact support. The amplitude gain of Hilbert's filter is 1 (except for $\nu = 0$, where the gain is zero), and its phase is

$$\beta(\nu) = \begin{cases} \pi/2 & \text{if } \nu > 0, \\ 0 & \text{if } \nu = 0, \\ -\pi/2 & \text{if } \nu < 0. \end{cases} \tag{11}$$

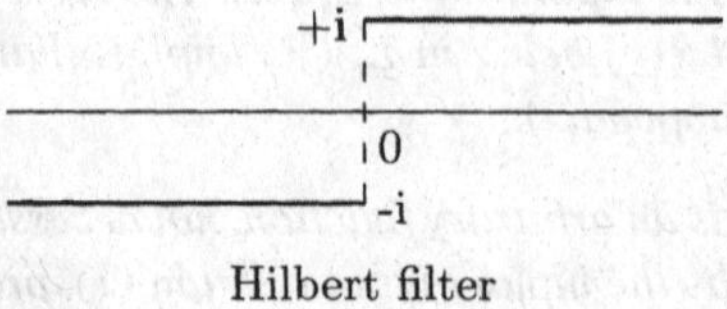

Figure B1.3. Hilbert frequency response

There is no function admitting the frequency response (10). There is, in fact, a generalized function (in the sense of the theory of distributions) with FT equal to $T(\nu)$. However, in signal theory, the Hilbert filter is used only in the theory of band-pass signals (see Section B1·2). For such signals the Hilbert filter coincides with a bona fide convolutional filter:

EXERCISE **B1.3.** *Show that the output $y(t)$ of the Hilbert filter, corresponding to a stable signal $x(t)$ having an FT $\hat{x}(\nu)$ that is null outside the frequency band $[-B, +B]$, can be expressed as*

$$y(t) = -\int_{\mathbb{R}} x(t-s)\frac{2\sin^2(\pi Bs)}{\pi s}\,ds.$$

Differentiation and Integration as Filters

Let $D(\mathcal{L})$ be the set of signals

$$x(t) = \int_{\mathbb{R}} \hat{x}(\nu)e^{2i\pi\nu t}\,d\nu, \tag{12}$$

where

$$\int_{\mathbb{R}} |\hat{x}(\nu)|\,d\nu < \infty \quad \text{and} \quad \int_{\mathbb{R}} |\nu|\,|\hat{x}(\nu)| < \infty.$$

Such signals are continuous and differentiable with derivative

$$\frac{d}{dt}x(t) = \int_{\mathbb{R}} (2i\pi\nu)\hat{x}(\nu)e^{2i\pi\nu t}\,d\nu. \tag{13}$$

(Apply the theorem of differentiation under the integral sign; 15 of the appendix). The mapping $x(t) \xrightarrow{\mathcal{L}} dx(t)/dt$ is a linear filter, called the *differentiating filter*, or differentiator, with frequency response

$$T(\nu) = 2i\pi\nu. \tag{14}$$

Let $D(\mathcal{L})$ be the set of signals of the form (12), where

$$\int_{\mathbb{R}} |\hat{x}(\nu)|\,d\nu < \infty \quad \text{and} \quad \int_{\mathbb{R}} \frac{|\hat{x}(\nu)|}{|\nu|}\,d\nu < \infty.$$

The signal

$$y(t) = \int_{\mathbb{R}} \frac{\hat{x}(\nu)}{2i\pi\nu}e^{2i\pi\nu t}\,d\nu$$

is in the domain of the preceding filter (the differentiator), and therefore,

$$\frac{d}{dt}y(t) = \int_{\mathbb{R}} \hat{x}(\nu)e^{2i\pi\nu t}\,d\nu = x(t).$$

The transformation $x(t) \xrightarrow{\mathcal{L}} y(t)$ is a homogeneous linear filter, which is called the *integrating filter*, or integrator, with frequency response

$$T(\nu) = \frac{1}{2i\pi\nu}. \tag{15}$$

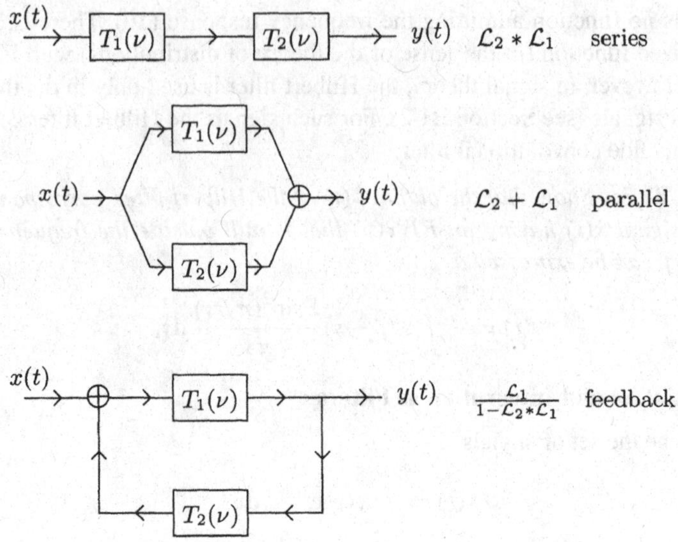

Figure B1.4. Series, parallel, and feedback configurations

Series, Parallel, and Feedback Configurations

We now describe the basic operations on filters (see Fig. B1.4). Let $\mathcal{L}_1$ and $\mathcal{L}_2$ be two convolutional filters with (stable) impulses responses $h_1(t)$ and $h_2(t)$ and frequency responses $T_1(\nu)$ and $T_2(\nu)$, respectively.

The *series* filter $\mathcal{L} = \mathcal{L}_2 * \mathcal{L}_1$ is, by definition, the convolutional filter with impulse response $h(t) = (h_1 * h_2)(t)$ and frequency response $T(\nu) = T_1(\nu)T_2(\nu)$. It operates as follows: The input $x(t)$ is first filtered by $\mathcal{L}_1$, and the output of $\mathcal{L}_1$ is then filtered by $\mathcal{L}_2$, to produce the final output $y(t)$.

The *parallel* filter $\mathcal{L} = \mathcal{L}_1 + \mathcal{L}_2$ is, by definition, the convolutional filter with impulse response $h(t) = h_1(t) + h_2(t)$ and frequency response $T(\nu) = T_1(\nu) + T_2(\nu)$. It operates as follows: The input $x(t)$ is filtered by $\mathcal{L}_1$, and "in parallel," it is filtered by $\mathcal{L}_2$, and the two outputs are added to produce the final output $y(t)$.

The *feedback* filter $\mathcal{L} = \mathcal{L}_1/(1 - \mathcal{L}_1 * \mathcal{L}_2)$ is, by definition, the convolutional filter with impulse response frequency response

$$T(\nu) = \frac{T_1(\nu)}{1 - T_1(\nu)T_2(\nu)}.$$

This filter will be a convolutional filter if and only if this frequency response is the FT of a stable impulse response. If this is not the case, one may define the feedback filter by the input–output relation

$$\hat{y}(\nu) = \frac{T_1(\nu)}{1 - T_1(\nu)T_2(\nu)}\hat{x}(\nu)$$

with, for instance, a definition along the lines of Example B1.2.

The filter $\mathcal{L}_1$ is the *forward loop* filter, whereas $\mathcal{L}_1$ is the *feedback loop* filter. The forward loop processes the total input, which consists of the input $x(t)$ plus the feedback input, that is, the output $y(t)$ processed by the feedback loop filter.

EXERCISE **B1.4.** *Consider the function*

$$T(\nu) = \frac{1}{1 + 4\pi^2\nu^2}.$$

Give the impulse response of the convolutional filter with the above frequency response $T(\nu)$. Interpret the filter as a feedback filter.

Filtering of Decomposable Signals

We introduce the notion of a decomposable signal, because it allows one to rewrite the results concerning Fourier transforms and Fourier series in a unified manner, without recourse to symbolic expressions in terms of the Dirac generalized functions or, more generally, to the theory of distributions.

DEFINITION **B1.4.** *The signal $s(t)$ is called decomposable if it can be put into the form*

$$s(t) = \int_{\mathbb{R}} e^{2i\pi\nu t}\,\hat{\mu}(d\nu), \tag{16}$$

where $\hat{\mu}$ is a complex measure on $\mathbb{R}$ whose total variation $|\hat{\mu}|$ is finite.

We recall that a complex measure of finite total variation is, by definition, a mapping $\mu : \mathcal{B}(\mathbb{R}) \mapsto \mathbb{C}$ ($\mathcal{B}(\mathbb{R})$ is the Borel sigma-field on $\mathbb{R}$; see the appendix) of the form

$$\mu = \mu_1 + i\mu_2,$$

where μ_1 and μ_2 are signed measures of finite total variation. A signed measure of finite total variation is a mapping $\mu : \mathcal{B}(\mathbb{R}) \mapsto \mathbb{R}$ of the form

$$\mu(C) = \mu_1(C \cap A) - \mu_2(C \cap \bar{A}),$$

for some $A \in \mathcal{B}(\mathbb{R})$ and all $C \in \mathcal{B}(\mathbb{R})$, where μ_1 and μ_2 are measures on $(\mathbb{R}, \mathcal{B}(\mathbb{R}))$ of finite total mass.

EXAMPLE **B1.4.** *Let $s(t)$ be a periodic signal with period T that is stable over its period and whose Fourier coefficients satisfy the condition*

$$\sum_{n\in\mathbb{Z}} |\hat{s}_n| < \infty.$$

Denote the Dirac measure at the point a by $\varepsilon_a(d\nu)$ and set

$$\hat{\mu}(d\nu) = \sum_{n\in\mathbb{Z}} \hat{s}_n \varepsilon_{\frac{n}{T}}(d\nu).$$

This measure is signed, has total variation $\sum_{n\in\mathbb{Z}} |\hat{s}_n| < \infty$, *and since*

$$\sum_{n\in\mathbb{Z}} \hat{s}_n e^{2i\pi \frac{n}{T} t} = \int_{\mathbb{R}} e^{2i\pi\nu t}\, \hat{\mu}(d\nu),$$

we see from the inversion formula that (16) *holds; therefore,* $s(t)$ *is decomposable.*

The measure $\hat{\mu}$ appearing in (16) will be called the *spectral decomposition* of the signal $s(t)$.

THEOREM **B1.1.** *Let* $x(t)$ *be a decomposable signal with spectral decomposition* $\hat{\mu}_x$*:*

$$x(t) = \int_{\mathbb{R}} e^{2i\pi\nu t}\, \hat{\mu}_x(d\nu), \tag{17}$$

and let $h(t)$ *be the impulse response, assumed stable, of a convolutional filter* $\mathcal{F}$.

The integral on the right-hand side of

$$y(t) = \int_{\mathbb{R}} h(t-s)x(s)\, ds$$

is well defined, and the spectral decomposition of $y(t)$ *is*

$$\hat{\mu}_y(d\nu) = T(\nu)\hat{\mu}_x(d\nu). \tag{18}$$

Proof:

$$\int_{\mathbb{R}}\int_{\mathbb{R}} |h(t-s)|\,|e^{2i\pi\nu s}|\,|\hat{\mu}_x|(d\nu)\, ds \le \left(\int_{\mathbb{R}} |h(t)|\, dt\right) |\hat{\mu}_x|(\mathbb{R}) < \infty,$$

and hence

$$\int_{\mathbb{R}} |h(t-s)|\,|x(s)|\, ds < \infty.$$

On the other hand,

$$\begin{aligned} y(t) &= \int_{\mathbb{R}} \left(h(t-s)\int_{\mathbb{R}} e^{2i\pi\nu s}\hat{\mu}_x(d\nu)\right) ds \\ &= \int_{\mathbb{R}} e^{2i\pi\nu(t-s)} \left(\int_{\mathbb{R}} e^{-2i\pi\nu(t-s)} h(t-s)\, ds\right) \hat{\mu}_x(d\nu), \end{aligned}$$

and hence

$$y(t) = \int_{\mathbb{R}} e^{2i\pi\nu t}\, T(\nu)\hat{\mu}_x(d\nu). \tag{19}$$

■

EXAMPLE **B1.5.** *In light of* (18), *one can interpret Eq.* (22) *of Theorem A2.1:*

$$\hat{y}_n = \hat{h}\left(\frac{n}{T}\right)\hat{x}_n,$$

where $h(t)$ *is a stable impulse response and* $x(t)$ *is a locally stable periodic signal with period* T*: If* $\sum_{n\in\mathbb{Z}} |\hat{x}_n| < \infty$, *then* $\sum_{n\in\mathbb{Z}} |\hat{y}_n| < \infty$, *since* $\hat{h}(\nu)$ *is bounded.*

The two signals $x(t)$ and $y(t)$ are therefore decomposable, and (19) *can be written*

$$\mu_y(d\nu) = \sum_{n\in\mathbb{Z}} \hat{y}_n \varepsilon_{\frac{n}{T}}(d\nu) = \sum_{n\in\mathbb{Z}} \hat{x}_n \hat{h}\left(\frac{n}{T}\right) \varepsilon_{\frac{n}{T}}(d\nu)$$

$$= \hat{h}(\nu)\mu_x(d\nu) = T(\nu)\hat{\mu}_x(d\nu).$$

Rational Filters as Differential Equations

The RC and LRC circuits are well-known filters, and they belong to the class of analog rational filters, which we proceed to define formally.

Let

$$P(z) = a_0 + \sum_{l=1}^{p} a_l z^l, \qquad Q(z) = b_0 + \sum_{l=1}^{q} b_l z^l \tag{20}$$

be two polynomials in the complex variable z. The coefficients a_p and b_q are nonzero, so that the degree of P is p, and the degree of Q is q. Moreover, we assume that $P(z)$ does *not* have purely imaginary roots:

$$P(z) \text{ has no roots in } i\mathbb{R}. \tag{21}$$

For all $\nu \in \mathbb{R}$, define

$$T(\nu) = \frac{Q(2i\pi\nu)}{P(2i\pi\nu)}. \tag{22}$$

We define a linear time-invariant filter $(\mathcal{L}, D(\mathcal{L}))$ as follows. First, we define the domain

$$D(\mathcal{L}) = \{x(t);\, x(t), \hat{x}(\nu), \text{ and } \nu^q \hat{x}(\nu) \in L^1_{\mathbb{C}}(\mathbb{R})\}. \tag{23}$$

We first observe that any function $x(t)$ in the domain is differentiable up to order q and that its jth derivative is

$$x^{(j)}(t) = \int_{\mathbb{R}} (2i\pi\nu)^j \hat{x}(\nu) e^{2i\pi\nu t}\, d\nu. \tag{24}$$

We now define the application itself:

$$\mathcal{L} : x(t) \to y(t) = \int_{\mathbb{R}} T(\nu)\hat{x}(\nu) e^{2i\pi\nu t}\, d\nu. \tag{25}$$

One has to verify that the integral in (25) is well defined. Indeed, $1/|P(2i\pi\nu)|$ is bounded because $P(z)$ has no imaginary root. In particular,

$$|T(\nu)||\hat{x}(\nu)| \le K|Q(2i\pi\nu)||\hat{x}(\nu)| \quad \text{for some } K < \infty.$$

Therefore, $|T(\nu)\hat{x}(\nu)|$ is integrable for all $x(t) \in D(\mathcal{L})$, and the integral in (25) is well defined for all such $x(t)$.

In fact, $T(\nu)\hat{x}(\nu)\nu^k$ is integrable for all k, $0 \le k \le p$. To check this, observe that $|\nu|^k/|P(2i\pi\nu)|$ is bounded for all $k \le p$ because $P(2i\pi\nu)$ is bounded away from zero ($P(z)$ has no imaginary root) and $|\nu|^k/|P(2i\pi\nu)|$ behaves as $|\nu|^{k-p}$ at ∞. Therefore, the output $y(t)$ is differentiable up to order p, and for $j \le p$,

$$y^{(j)}(t) = \int_{\mathbb{R}} (2i\pi\nu)^j T(\nu)\hat{x}(\nu)e^{2i\pi\nu t}\,d\nu. \tag{26}$$

From (24), (26), and (22), it follows that the input $x(t)$ and the output $y(t)$ are linked by the differential equation

$$a_0 y(t) + \sum_{l=1}^{p} a_l y^{(l)}(t) = b_0 x(t) + \sum_{l=1}^{q} b_l x^{(l)}(t),$$

that is, symbolically,

$$P\left(\frac{\partial}{\partial t}\right) y(t) = Q\left(\frac{\partial}{\partial t}\right) x(t). \tag{27}$$

This is the time-domain relation corresponding to the frequency-domain relation

$$P(2i\pi\nu)\hat{y}(\nu) = Q(2i\pi\nu)\hat{x}(\nu).$$

Differential Equations as Rational Filters

We now consider the inverse problem: Let $x(t) \in D(\mathcal{L})$—in particular, $x(t)$ is differentiable up to order q—and let $y(t)$ be a solution of the differential equation (27). Is it possible to express this solution as

$$\int_{\mathbb{R}} \frac{Q(2i\pi\nu)}{P(2i\pi\nu)} \hat{x}(\nu)e^{2i\pi\nu t}\,d\nu\ ?$$

The answer is "no, in general" and "yes, asymptotically" if we impose the following condition:

$$P(z) \text{ is strictly stable,} \tag{28}$$

that is, the real parts of all the roots of $P(z)$ are strictly negative. Then, $y(t), t \ge t_0$, the solution of (27) with arbitrary initial conditions $y(t_0) = y_0$, $y_0^{(j)}(t_0) = y_0^{(j)}$ $(1 \le j \le p-1)$, satisfies

$$\lim_{t\uparrow\infty} \left(y(t) - \int_{\mathbb{R}} \frac{Q(2i\pi\nu)}{P(2i\pi\nu)} \hat{x}(\nu)e^{2i\pi\nu t}\,d\nu \right) = 0. \tag{29}$$

Proof: The general solution of (27) is the sum of a particular solution of (27) and of the general solution of the differential equation without a right-hand side,

$$P\left(\frac{\partial}{\partial t}\right) y(t) = 0. \tag{30}$$

Therefore, since

$$\int_{\mathbb{R}} \frac{Q(2i\pi\nu)}{P(2i\pi\nu)} \hat{x}(\nu)e^{2i\pi\nu t}$$

is a particular solution of (27), we have to show that $\lim_{t\uparrow\infty} z(t) = 0$ for the general solution of (30). This follows from the theory of linear differential equations because the characteristic polynomial $P(z)$ of (30) has all its roots in the open left half complex plane (see [B5]). ■

If $P(z)$ is not strictly stable, there are initial conditions such that

$$y(t) - \int_{\mathbb{R}} \frac{Q(2i\pi\nu)}{P(2i\pi\nu)} \hat{x}(\nu) e^{2i\pi\nu t}\, d\nu$$

does not tend to 0 as $t \to \infty$.

EXAMPLE **B1.6.** *Consider the LRC circuit (see Fig. B1.5). Its input and output are related through the differential equation*

$$LC\ddot{y}(T) + RC\dot{y}(t) + y(t) = x(t),$$

where $\dot{y}(t)$ and $\ddot{y}(t)$ are the first and second derivatives of $y(t)$. The roots of the characteristic polynomial

$$P(z) = 1 + RCz + LCz^2$$

are given by the formula

$$z = \frac{-R \pm \sqrt{R^2 - 4L/C}}{2L},$$

and their real parts are always strictly negative. Therefore, the system is strictly stable, and the permanent regime when the input is $x(t) \in D(\mathcal{L})$ is

$$y(t) = \int_{\mathbb{R}} \frac{1}{1 + RC(2i\pi\nu) + LC(2i\pi\nu)^2} \hat{x}(\nu) e^{2i\pi\nu t}\, d\nu.$$

We note that $Q(z) \equiv 1$ in this example, and therefore,

$$D(\mathcal{L}) = \{x(t) : x(t), \hat{x}(\nu) \in L^1_{\mathbb{C}}(\mathbb{R})\}.$$

Rational Filters as Convolutional Filters

Going back to the general case described by Eqs. (20)–(25), we pose the problem: Is the filter of convolutional type? The answer is yes if and only if $q < p$. Indeed, consider the factorization of $P(z)$:

$$P(z) = a_p \prod_{k=1}^{r} (z - z_k)^{m_k},$$

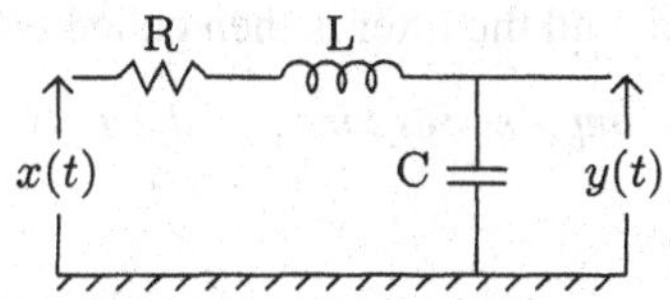

Figure B1.5. The LRC circuit

where z_k is a root of order m_k. We have the decomposition

$$\frac{Q(z)}{P(z)} = \sum_{k=0}^{q-p} \alpha_k z^k + \sum_{k=1}^{r} \sum_{j=1}^{m_k} \frac{\beta_{k_j}}{(z - z_k)^j},$$

and therefore,

$$T(\nu) = \sum_{k=0}^{q-p} \alpha_k (2i\pi\nu)^k + \sum_{k=1}^{r} \sum_{j=1}^{m_k} \frac{\beta_{k_j}}{(2i\pi\nu - z_k)^j}.$$

If $\operatorname{Re}(z_k) > 0$,

$$\int_{-\infty}^{0} \frac{t^{j-1}}{(j-1)!} e^{z_k t} e^{-2i\pi\nu t}\, \mathrm{d}t = -\frac{1}{(2i\pi\nu - z_k)^j}.$$

If $\operatorname{Re}(z_k) < 0$,

$$\int_{0}^{\infty} \frac{t^{j-1}}{(j-1)!} e^{z_k t} e^{-2i\pi\nu t}\, \mathrm{d}t = +\frac{1}{(2i\pi\nu - z_k)^j}.$$

(Remember that the case $\operatorname{Re}(z_k) = 0$ has been excluded.) Defining

$$h(t) = \begin{cases} -\displaystyle\sum_{k;\operatorname{Re}(z_k)>0} \sum_{j=1}^{m_k} \frac{\beta_{k_j} t^{j-1}}{(j-1)!} e^{z_k t} & \text{for } t < 0, \\ \displaystyle\sum_{k;\operatorname{Re}(z_k)<0} \sum_{j=1}^{m_k} \frac{\beta_{k_j} t^{j-1}}{(j-1)!} e^{z_k t} & \text{for } t \geq 0, \end{cases} \tag{31}$$

we have

$$h(t) \quad \overset{\text{FT}}{\rightarrow} \quad \sum_{k=1}^{r} \sum_{j=1}^{m_k} \frac{\beta_{k_j}}{(2i\pi\nu - z_k)^j}.$$

The input–output mapping is therefore

$$y(t) = \sum_{k=0}^{q-p} \alpha_k x^{(k)}(t) + \int_{\mathbb{R}} h(t-s)x(s)\, \mathrm{d}s. \tag{32}$$

In particular,

$$y(t) = \int_{\mathbb{R}} h(t-s)x(s)\, \mathrm{d}s \tag{33}$$

if and only if $q < p$. If, moreover, $P(z)$ is strictly stable (no roots in the closed right half-plane), then, as the expression (31) of the impulse response shows, the impulse response is causal, and the filter is then called realizable.

EXERCISE **B1.5.** *Give the impulse response of the LRC filter in the case $R^2 < 4L/C$.*

We observe that the input–output relationship (32) is meaningful for all $x(t) \in D(\mathcal{L}')$, where $D(\mathcal{L}')$ consists of the stable complex signals that are differentiable up to order $\max(0, q - p)$. Therefore, one can consider that the filter $(\mathcal{L}', D(\mathcal{L}'))$, where $\mathcal{L}'$ is described by (32), is an extension of the original filter $(\mathcal{L}, D(\mathcal{L}))$. We may consider that (32) is an extension of the differential equation (27). For some functions of the extended domain, the input–output relationship is not a differential equation.

EXERCISE **B1.6.** *Give the extended filter corresponding to the differential equation*

$$y'' - \frac{1}{2}y = x'' - \frac{1}{3}x.$$

Butterworth Filters

We consider the problem of implementing an approximation of the ideal low-pass (B) filter (with cut-off frequency B) by means of a *stable* and *realizable* filter with *real* impulse response $h(t)$. Let $T(\nu)$ be the frequency response of the approximating filter. The following family of filters, called *Butterworth filters*, has been proposed:

$$|T(\nu)|^2 = \frac{1}{1 + \left(\frac{\nu}{B}\right)^{2n}}. \tag{34}$$

(As $n \to \infty$, the filter looks more and more like an ideal low-pass filter.) One seeks $T(\nu)$ of the form

$$T(\nu) = \frac{K}{P(2i\pi\nu)},$$

where $P(z)$ has all its roots strictly to the left of the imaginary axis in order to guarantee stability and causality.

The roots of the polynomial $1 + (\nu/B)^{2n}$ are

$$\nu_\ell = B\, e^{i\frac{\pi}{2n}(2\ell+1)}, \qquad 0 \le \ell \le 2n + 1.$$

We reorder these roots in such a way that $\nu_1, \nu_1^*, \ldots, \nu_n, \nu_n^*$, are the $2n$ roots, where $\nu_1, \ldots, \nu_n$ have strictly positive imaginary parts. We shall allocate $\nu_1, \ldots, \nu_n$ to $T(\nu)$, thus proposing

$$T(\nu) = \frac{B^n}{\prod_{j=1}^{n}(\nu - \nu_j)}. \tag{35}$$

This is the frequency response of a *real* filter (i.e., $T^*(\nu) = T(-\nu)$) because any root among $\nu_1, \ldots, \nu_n$ is purely imaginary or it can be associated with another root symmetric with respect to the imaginary axis (see Fig. B1.6).

In the case $n = 2$, we find

$$T(\nu) = \frac{\nu_1\nu_2}{(\nu - \nu_1)(\nu - \nu_2)},$$

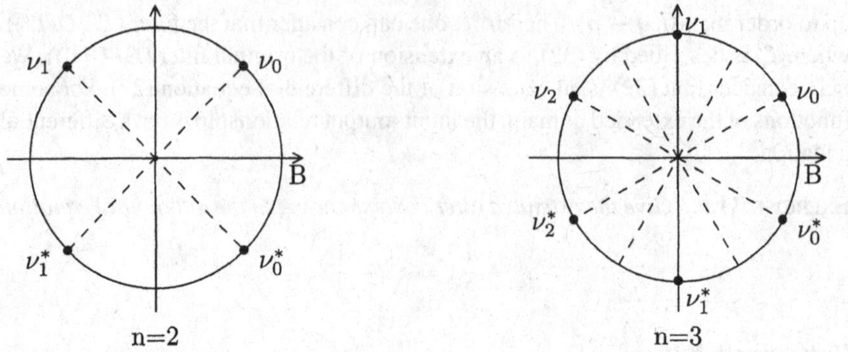

Figure B1.6. Roots of $1 + (\nu/B)^{2n}$ for $n = 2$ and $n = 3$

and in the case $n = 3$,

$$T(\nu) = \frac{-\nu_1\nu_2\nu_3}{(\nu - \nu_1)(\nu - \nu_2)(\nu - \nu_3)}.$$

EXERCISE **B1.7.** *Show that the Butterworth filter of order $n = 2$ can be implemented by an LRC circuit.*

B1·2 Band-Pass Signals

In this section, we give the basic facts concerning frequency transposition and study the phenomena associated with it, such as cross-talk in quadrature multiplexing, and channel dispersion.

Complex Envelope

The first relevant notion is that of a base-band signal.

DEFINITION **B1.1.** *A band-pass (ν_0, B) signal, where $B < \nu_0$, is a stable signal whose FT is null if $|\nu| \notin [-B + \nu_0, \nu_0 + B]$. A base-band (B) signal is a stable signal $s(t)$ whose FT is null outside the interval $[-B, +B]$.*

It will be assumed, moreover, that $s(t)$ is real and hence that its FT is Hermitian even:

$$\hat{s}(-\nu) = \hat{s}(\nu)^*.$$

We are going to show that a real band-pass signal $s(t)$ has the representation

$$s(t) = m(t)\cos(2\pi\nu_0 t) - n(t)\sin(2\pi\nu_0 t), \tag{36}$$

where $m(t)$ and $n(t)$ are two signals that are *real*, and *base-band* (B). The base-band signals $m(t)$ and $n(t)$ are the *quadrature components* of the band-pass signal $s(t)$.

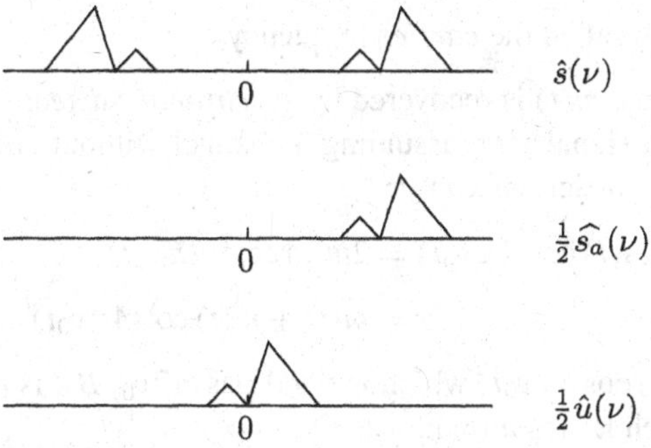

Figure B1.7. Complex envelope in the Fourier domain

One way of proving (36) is to form the *analytic signal of* $s(t)$

$$s_a(t) = 2\int_0^\infty \hat{s}(\nu)e^{2i\pi\nu t}\,d\nu, \tag{37}$$

and then its *complex envelope* $u(t)$ (see Fig. B1.7)

$$u(t) = \int_{\mathbb{R}} \hat{s}_a(\nu+\nu_0)e^{2i\pi\nu t}\,d\nu. \tag{38}$$

EXERCISE **B1.8.** *Show that the FT of the signal* $\text{Re}\,\{u(t)e^{2i\pi\nu_0 t}\}$ *is* $\hat{s}(\nu)$, *and thus*

$$s(t) = \text{Re}\,\{u(t)e^{2i\pi\nu_0 t}\}. \tag{39}$$

Let $m(t)$ and $n(t)$ be the real and imaginary parts of $u(t)$:

$$u(t) = m(t) + in(t). \tag{40}$$

The quadrature decomposition (36) follows from (39) and (40).

EXERCISE **B1.9.** *Show that*

$$\widehat{m}(\nu) = \frac{\hat{u}(\nu) + \hat{u}(-\nu)^*}{2}, \tag{41a}$$

$$\hat{n}(\nu) = \frac{\hat{u}(\nu) - \hat{u}(-\nu)^*}{2i} \tag{41b}$$

and that

$$\widehat{m}(\nu) = \{\hat{s}(\nu+\nu_0) + \hat{s}(\nu-\nu_0)\}1_{[-B,+B]}(\nu), \tag{42a}$$

$$\hat{n}(\nu) = -i\{\hat{s}(\nu+\nu_0) - \hat{s}(\nu-\nu_0)\}1_{[-B,+B]}(\nu). \tag{42b}$$

Frequency Transposition and Quadrature Multiplexing

Frequency transposition is the operation that transforms a real signal $m(t)$, base-band (B), into the band-pass (ν_0, B) signal

$$s(t) = m(t)\cos 2\pi\nu_0 t.$$

The frequency ν_0 is called the *carrier* frequency.

The original signal $m(t)$ is recovered by *synchronous detection*: One first multiplies the received signal $s(t)$ (assuming a channel without noise, distortion, or attenuation) by the carrier $\cos 2\pi\nu_0 t$:

$$\begin{aligned}2s(t)\cos(2\pi\nu_0 t) &= 2m(t)\cos^2(2\pi\nu_0 t)\\ &= m(t) + m(t)\cos(4\pi\nu_0 t),\end{aligned}$$

and the signal $m(t)\cos 4\pi\nu_0 t$, which is band-pass $(2\nu_0, B)$, is eliminated by the low-pass (B), which leaves $m(t)$ intact.

Since a real signal such as $m(t)$ has a Hermitian symmetric FT, the frequency transposition technique uses a bandwidth $2B$, and therefore, there is a waste of bandwidth: One should be able to transmit *two* real signals in the base-band (B) on a bandwidth of $2B$. There are several ways of doing this. One of them is *quadrature multiplexing* (quadrature amplitude modulation, or QAM). In this technique, in order to transmit two real base-band (B) signals $m(t)$ and $n(t)$, one sends the signal $s(t) = m(t)\cos(2\pi\nu_0 t) - n(t)\sin(2\pi\nu_0 t)$.

EXERCISE **B1.10.** *Let $s(t)$ be a base-band (B) signal of finite energy. What is the support of the FT of the signal $s(t)^2$?*

EXERCISE **B1.11.** *Show that, in order to recover $m(t)$ (resp., $n(t)$), one can multiply $s(t)$ by $2\cos(2\pi\nu_0 t)$ (resp., $2\sin(2\pi\nu_0 t)$) and then pass the resulting signal through a low-pass (B) (see Fig. B1.8).*

Band-Pass Filtering

When the band-pass signal (36) is passed through a filter with frequency response $T(\nu)$, we may, without loss of generality, consider that $T(\nu) = 0$ if $|\nu| \notin [\nu_0 - B, \nu_0 + B]$, since filtering is expressed in the frequency domain by multiplication of the FTs. Hence it will be assumed that the impulse response $h(t)$ of the filter is also a band-pass (ν_0, B) function.

The output signal $y(t)$ has as FT

$$\hat{y}(\nu) = T(\nu)\hat{s}(\nu), \tag{43}$$

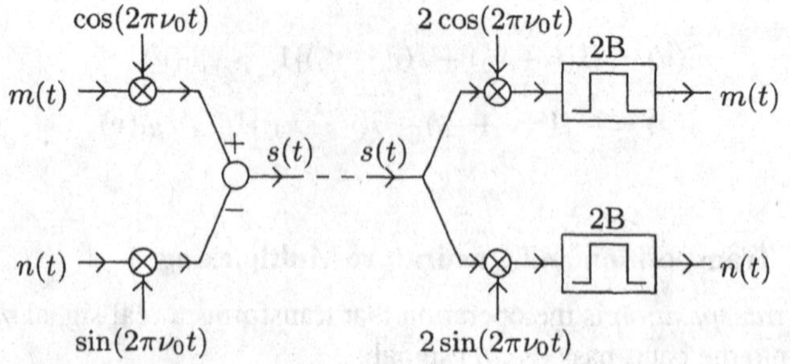

Figure B1.8. Quadrature multiplexing

and it is therefore also band-pass (ν_0, B).

EXERCISE **B1.12.** *Show that if we denote by* $v(t)$ *and* $u(t)$ *the complex envelopes of* $y(t)$ *and* $s(t)$*, respectively, then*

$$\hat{v}(\nu) = T(\nu + \nu_0)\hat{u}(\nu). \tag{44}$$

The equality (36) *is a base-band representation of the band-pass filtering equality* (43).

We shall describe two effects that are specific of frequency transposition. The first one is the phenomenon of cross-talk in quadrature multiplexed channels.

Cross-Talk

Suppose we use quadrature multiplexing; we thus send two band-pass messages $m(t)$ and $n(t)$ in the form

$$s(t) = m(t)\cos(2\pi\nu_0 t) - n(t)\sin(2\pi\nu_0 t).$$

Ideal reception (without distortion in the channel) is performed by synchronous detection whereby $m(t)$ and $n(t)$ are recovered. Their FTs are given by (42a) and (42b), respectively.

Suppose there is distortion in the channel and that, consequently, the received signal is $s'(t)$. We then obtain, after synchronous detection, $m'(t)$ and $n'(t)$ with respective FTs

$$\widehat{m}'(\nu) = \{\hat{s}'(\nu + \nu_0) + \hat{s}'(\nu - \nu_0)\}1_{[-B,+B]}(\nu),$$

$$\hat{n}'(\nu) = -i\{\hat{s}'(\nu + \nu_0) - \hat{s}'(\nu - \nu_0)\}1_{[-B,+B]}(\nu).$$

Let us assume that the distortion $s(t) \to s'(t)$ is a linear filtering with frequency response $T(\nu)$. Let us note that $T(\nu)$ is Hermitian symmetric, as it is the FT of a real impulse response $h(t)$. We have $\hat{s}'(\nu) = \hat{s}(\nu)T(\nu)$, and therefore,

$$\widehat{m}'(\nu) = \{T(\nu + \nu_0)\hat{s}(\nu + \nu_0) + T(\nu - \nu_0)\hat{s}(\nu - \nu_0)\}1_{[-B,+B]}(\nu),$$

$$\hat{n}'(\nu) = -i\{T(\nu + \nu_0)\hat{s}(\nu + \nu_0) - T(\nu - \nu_0)\hat{s}(\nu - \nu_0)\}1_{[-B,+B]}(\nu).$$

It appears that $\widehat{m}'(\nu)$ in general cannot be expressed as a function of $\widehat{m}(\nu)$ alone. It depends on both $\widehat{m}(\nu)$ and $\hat{n}(\nu)$, and therefore, in general, there is *interference* between the two paths. However, under the condition that $T(\nu)$ be Hermitian symmetric about ν_0 in the band of width $2B$ centered on ν_0, that is,

$$T(\nu + \nu_0) = T^*(\nu_0 - \nu) \quad \text{for all } \nu \in [-B, +B], \tag{45}$$

or, again, in view of the Hermitian symmetry of $T(\nu)$ about 0,

$$T(\nu + \nu_0) = T(\nu - \nu_0) \quad \text{for all } \nu \in [-B, +B], \tag{46}$$

then

$$\widehat{m}'(\nu) = G(\nu)\widehat{m}(\nu), \qquad \hat{n}'(\nu) = G(\nu)\hat{n}(\nu), \tag{47}$$

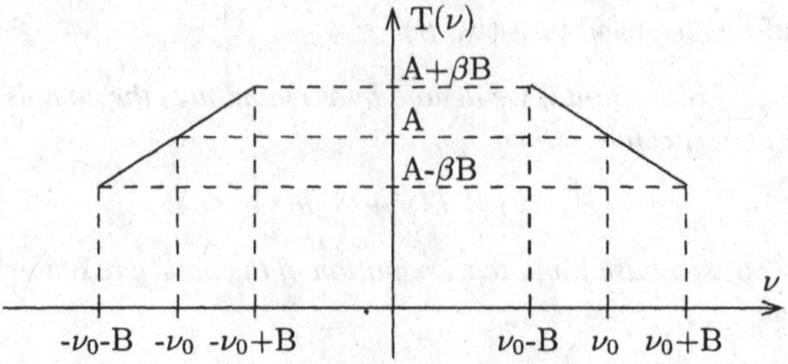

Figure B1.9. The frequency response in Exercise B1.13

where

$$G(\nu) = T(\nu + \nu_0). \tag{48}$$

In this case there is only a linear distortion, represented by independent filtering of $m(t)$ and $n(t)$. After identification of the channel (that is, identification of $T(\nu)$), we can therefore recover the signals in the two paths.

EXERCISE **B1.13.** *Suppose that in the band* $[\nu_0 - B, \nu_0 + B]$, $T(\nu)$ *has the following form (see Fig. B1.9):*

$$T(\nu + \nu_0) = A + \beta\nu, \qquad T(\nu - \nu_0) = A - \beta\nu.$$

Show that $m'(t)$ *is a linear combination of* $m(t)$ *and* $(\mathrm{d}/\mathrm{d}t)\, n(t)$.

We shall now study another phenomenon associated with frequency transposition, that of group delay.

Dispersive Channels

A *dispersive channel* is a homogeneous linear filter with frequency response

$$T(\nu) = Ke^{i\beta(\nu)}, \tag{49}$$

where K is a complex constant that will be taken equal to unity. This channel transforms the complex sinusoid $e^{2i\pi\nu t}$ into the delayed complex sinusoid $e^{i(2\pi\nu t + \beta(\nu))}$, where $\beta(\nu)$ is the phase of the filter at the frequency ν.

Let $s(t)$ be a real signal, band-pass (ν_0, B), of the form $s(t) = m(t)\cos 2\pi\nu_0 t$. Let $y(t)$ be the signal obtained by passing $s(t)$ through the dispersive channel. The corresponding base-band equivalent filter has the frequency representation

$$\hat{v}(\nu) = T(\nu + \nu_0)\widehat{m}(\nu),$$

where $\hat{v}(\nu)$ is the FT of the complex envelope $v(t)$ of $y(t)$ [see (43)].

Suppose that in the band $[\nu_0 - B, \nu_0 + B]$, the dispersion has a first-order expansion

$$\beta(\nu + \nu_0) \simeq \beta(\nu_0) + \nu \left.\frac{\partial\beta}{\partial\nu}\right|_{\nu=\nu_0}, \qquad \nu \in [-B, +B];$$

then (approximately)

$$\hat{v}(\nu) = \widehat{m}(\nu)e^{-2i\pi\nu_0\tau_p - 2i\pi\nu\tau_g},$$

where

$$\tau_p = -\frac{\beta(\nu_0)}{2\pi\nu_0} \tag{50}$$

and

$$\tau_g = -\frac{1}{2\pi}\frac{\partial\beta}{\partial\nu}\bigg|_{\nu=\nu_0}. \tag{51}$$

Therefore, we have

$$v(t) = m(t - \tau_g)e^{-2i\pi\nu_0\tau_p}.$$

Now,

$$y(t) = \text{Re}\,\{v(t)e^{2i\pi\nu_0 t}\}.$$

Hence we have

$$y(t) = m(t - \tau_g)\cos 2\pi\nu_0(t - \tau_p). \tag{52}$$

The constants τ_p and τ_g are the *phase delay* and *group delay*, respectively.

B2
Sampling

B2·1 Reconstruction and Aliasing

In a digital communication system, an analog signal $\{s(t)\}_{t\in\mathbb{R}}$ must be transformed into a sequence of binary symbols, 0 and 1. This binary sequence is generated by first sampling the analog signal, that is, extracting a sequence of samples $\{s(n\Delta)\}_{n\in\mathbb{Z}}$, and then quantizing which means converting each sample into a block of 0 and 1.

The first question that arises is: To what extent does the sample sequence represent the original signal? This cannot be true without further assumptions since obviously an infinite number of signals fit a given sequence of samples.

The second question is: How do we efficiently reconstruct the signal from its samples?

The Shannon–Nyquist Theorem

We begin with a general result that will then be applied to the study of undersampling and both oversampling.

THEOREM **B2.1.** *Let $s(t)$ be a stable and continuous complex signal with Fourier transform $\hat{s}(\nu) \in L^1_{\mathbb{C}}(\mathbb{R})$, and assume in addition that, for some $0 < B < \infty$,*

$$\sum_{n\in\mathbb{Z}} \left| s\left(\frac{n}{2B}\right)\right| < \infty. \tag{53}$$

Then

$$\sum_{j\in\mathbb{Z}} \hat{s}(\nu + j2B) = \frac{1}{2B}\sum_{n\in\mathbb{Z}} s\left(\frac{n}{2B}\right) e^{-2i\pi\nu\frac{n}{2B}}, \text{ a.e.} \tag{54}$$

Let $h(t)$ be a complex signal of the form

$$h(t) = \int_R T(\nu) e^{2i\pi\nu t}\, d\nu, \tag{55}$$

where $T(\nu) \in L^1_{\mathbb{C}}(\mathbb{R})$. The signal

$$\tilde{s}(t) = \frac{1}{2B}\sum_{n\in\mathbb{Z}} s\left(\frac{n}{2B}\right) h\left(t - \frac{n}{2B}\right) \tag{56}$$

then admits the representation

$$\tilde{s}(t) = \int_{\mathbb{R}} \left\{\sum_{j\in\mathbb{Z}} \hat{s}(\nu + j2B)\right\} T(\nu) e^{2i\pi\nu t}\, d\nu. \tag{57}$$

Proof: By Theorem A2.3, the $2B$-periodic function $\Phi(\nu) = \sum_{j\in\mathbb{Z}} \hat{s}(\nu + j2B)$ is locally integrable, and its nth Fourier coefficient is

$$\frac{1}{2B}\int_{\mathbb{R}} \hat{s}(\nu) e^{-2i\pi\frac{n}{2B}\nu}\, d\nu,$$

that is, since the Fourier inversion formula for $s(t)$ holds ($\hat{s}(\nu)$ is integrable) and it holds *everywhere* ($s(t)$ is continuous), the nth Fourier coefficient of $\Phi(\nu)$ is in fact equal to

$$s\left(-\frac{n}{2B}\right).$$

The formal Fourier series of $\Phi(\nu)$ is therefore

$$\frac{1}{2B}\sum_{n\in\mathbb{Z}} s\left(\frac{n}{2B}\right) e^{-2i\pi\frac{n}{2B}\nu}.$$

In view of condition (53), the Fourier inversion formula holds a.e. (Theorem A2.2), that is, $\Phi(\nu)$ is almost everywhere equal to its Fourier series. This proves (54).

Since the frequency response $T(\nu) \in L^1_{\mathbb{C}}(\mathbb{R})$, the impulse response $h(t)$ given by (55) is bounded and uniformly continuous, and therefore $\tilde{s}(t)$ is bounded and continuous (the right-hand side of (56) is a normally convergent series—by (53)— of bounded and continuous functions). Also, upon substituting (55) in (56), we obtain

$$\tilde{s}(t) = \frac{1}{2B}\sum_{n\in\mathbb{Z}} s\left(\frac{n}{2B}\right)\int_{\mathbb{R}} T(\nu)\, e^{2i\pi\nu(t-\frac{n}{2B})}\, d\nu$$

$$= \int_{\mathbb{R}} \left\{\sum_{n\in\mathbb{Z}} \frac{1}{2B} s\left(\frac{n}{2B}\right) e^{-2i\pi\nu\frac{n}{2B}}\right\} T(\nu)\, e^{2i\pi\nu t}\, d\nu.$$

(The interchange of integration and summation is justified by Fubini's theorem because

$$\int_{\mathbb{R}} \sum_{n\in\mathbb{Z}} \left| s\left(\frac{n}{2B}\right)\right| |T(\nu)| \,\mathrm{d}\nu = \left(\sum_{n\in\mathbb{Z}} \left| s\left(\frac{n}{2B}\right)\right|\right)\left(\int_{\mathbb{R}} |T(\nu)| \,\mathrm{d}\nu\right) < \infty.)$$

Therefore,

$$\tilde{s}(t) = \int_{\mathbb{R}} g(\nu) e^{2i\pi\nu t} \,\mathrm{d}\nu,$$

where

$$g(\nu) = \frac{1}{2B}\left\{\sum_{n\in\mathbb{Z}} s\left(\frac{n}{2B}\right) e^{-2i\pi\nu\frac{n}{2B}}\right\} T(\nu).$$

The result (57) then follows from (54). ■

We now state the Shannon–Nyquist sampling theorem.

THEOREM **B2.2.** *Let $s(t)$ be a stable and continuous signal whose FT $\hat{s}(\nu)$ vanishes outside $[-B, +B]$, and assume condition* (53) *is satisfied. We can then recover $s(t)$ from its samples $s(n/2B)$, $n \in \mathbb{Z}$, by the formula*

$$s(t) = \sum_{n\in\mathbb{Z}} s\left(\frac{n}{2B}\right) \operatorname{sinc}(2Bt - n), \ a.e. \tag{58}$$

Proof: This is a direct consequence of the previous theorem, with $T(\nu)$ the frequency response of the low-pass (B). Indeed,

$$\left\{\sum_{j\in\mathbb{Z}} \hat{s}(\nu + j2B)\right\} T(\nu) = \hat{s}(\nu) 1_{[-B,+B]}(\nu) = \hat{s}(\nu),$$

and therefore, by (57),

$$\tilde{s}(t) = \int_{\mathbb{R}} \hat{s}(\nu) e^{2i\pi\nu t} \,\mathrm{d}\nu = s(t).$$

The second equality is an almost everywhere equality; it holds everywhere when $s(t)$ is a continuous signal (see Corollary A1.2). ■

If we interpret $s(n/2B)h(t - n/2B)$ as the response of the low-pass (B) when a Dirac impulse of height $s(n/2B)$ is applied at time $n/2B$, the right-hand side of Eq. (58) is the response of the low-pass (B) to the Dirac comb (see Figs. B2.1 and B2.2)

$$s_i(t) = \frac{1}{2B} \sum_{n\in\mathbb{Z}} s\left(\frac{n}{2B}\right) \delta\left(t - \frac{n}{2B}\right). \tag{59}$$

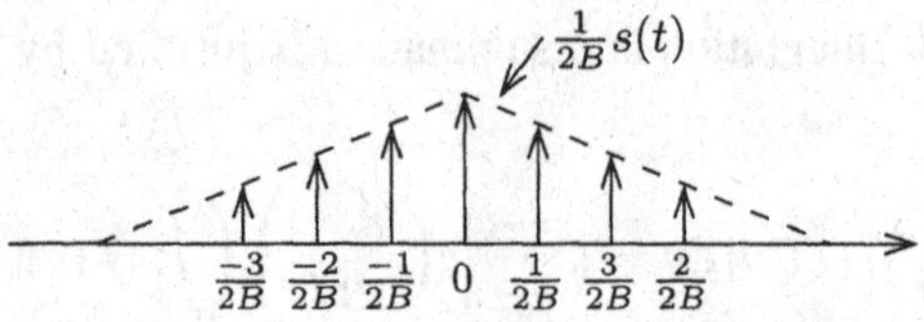

Figure B2.1. The Dirac comb of (59)

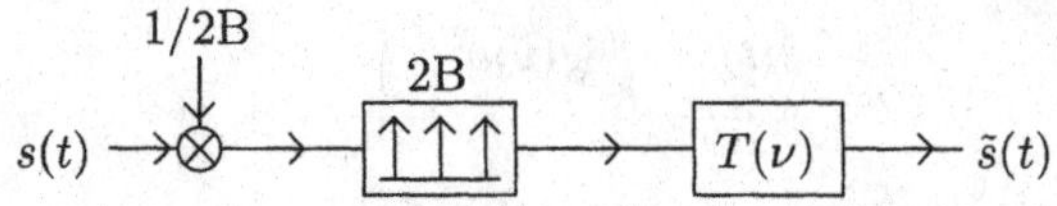

Figure B2.2. Sampling and reconstruction

Sample and Hold

In practice, the Dirac comb is replaced by a train of "true" functions. Instead of the above train of impulses one of the techniques of reconstruction (called *sample and hold*) uses the train of rectangles

$$s_{i,\tau}(t) = \frac{1}{2B}\sum_{n\in\mathbb{Z}} s\left(\frac{n}{2B}\right) g_\tau\left(t - \frac{n}{2B}\right),$$

where $g_\tau(t)$ is a rectangle of base τ and unit area,

$$g_\tau(t) = \frac{1}{\tau}\, 1_{[0,\tau]}(t),$$

an approximation of the Dirac impulse as τ becomes large. This signal is then filtered by a low-pass (B), to produce the signal $s_\tau(t)$. We show that the result is a smoothed version of the original signal:

$$s_\tau(t) = \frac{1}{\tau}\int_0^\tau s(t-u)\,du.$$

(Observe that we cannot use Theorem B2.1 as such; why?). Condition (53) implies that $s_{i,\tau}(t)$ is integrable and has an FT given by

$$\widehat{s_{i,\tau}}(\nu) = \frac{1}{2B}\sum_{n\in\mathbb{Z}}\left(\frac{n}{2B}\right)\widehat{g_\tau}(\nu)e^{-2i\pi\frac{n}{2B}}$$

$$= \widehat{g_\tau}(\nu)\left(\frac{1}{2B}\sum_{n\in\mathbb{Z}} s\left(\frac{n}{2B}\right)e^{-2i\pi\frac{n}{2B}}\right).$$

The signal $s_\tau(t)$ is obtained by low-pass (B) filtering of the stable signal $s_{i,\tau}(t)$. Since the impulse response of a low-pass is not integrable, we cannot use the current version of the convolution–multiplication rule as it is. However, we shall proceed formally because the result is justified by a more appropriate version of the convolution–multiplication rule (Theorem C3.4). We therefore have

$$\widehat{s_T}(\nu) = \widehat{g_T}(\nu)\left(\frac{1}{2B}\sum_{n\in\mathbb{Z}} s\left(\frac{n}{2B}\right)e^{-2i\pi\frac{n}{2B}}\,1_{[-B,+B]}(\nu)\right) = \widehat{g_T}(\nu)\hat{s}(\nu).$$

The result then follows by the inversion formula and the convolution–multiplication formula (the current version, this time).

Aliasing

What happens in the Shannon–Nyquist sampling theorem if one supposes that the signal is base-band (B), although it is not the case in reality?

Suppose that a stable signal $s(t)$ is sampled at frequency $2B$ and that the resulting impulse train is applied to the low-pass (B) with impulse response $h(t) = 2B\,\mathrm{sinc}\,(2Bt)$, to obtain, after division by $2B$, the signal

$$\tilde{s}(t) = \sum_{n\in\mathbb{Z}} s\left(\frac{n}{2B}\right)\mathrm{sinc}\,(2Bt - n).$$

What is the FT of this signal? The answer is given by the following theorem, which is a direct consequence of Theorem B2.1.

THEOREM **B2.3.** *Let $s(t)$ be a stable and continuous signal such that condition* (53) *is satisfied. The signal*

$$\tilde{s}(t) = \sum_{n\in\mathbb{Z}} s\left(\frac{n}{2B}\right)\mathrm{sinc}\,(2Bt - n) \tag{60}$$

admits the representation

$$\tilde{s}(t) = \int_{\mathbb{R}} \hat{\tilde{s}}(\nu)e^{2i\pi\nu t}\,d\nu,$$

where

$$\hat{\tilde{s}}(\nu) = \left\{\sum_{k\in\mathbb{Z}} \hat{s}(\nu + j2B)\right\}1_{[-B,+B]}(\nu). \tag{61}$$

If $\tilde{s}(t)$ is integrable, then $\hat{\tilde{s}}(\nu)$ is its FT, by the Fourier inversion theorem. This FT is obtained by superposing, in the frequency band $[-B, +B]$, the translates by multiples of $2B$ of the initial spectrum $\hat{s}(\nu)$. This superposition constitutes the phenomenon of *spectrum folding*, and the distortion that it creates is called *aliasing* (see Fig. B2.3).

EXERCISE **B2.1.** *Show that if the signal*

$$s(t) = \left(\frac{\sin(2\pi Bt)}{\pi t}\right)^2$$

is sampled at rate $1/2B$ and if the resulting train of impulses is filtered by a low-pass (B) and divided by $2B$, the result is the signal

$$\frac{\sin(2\pi Bt)}{\pi t}.$$

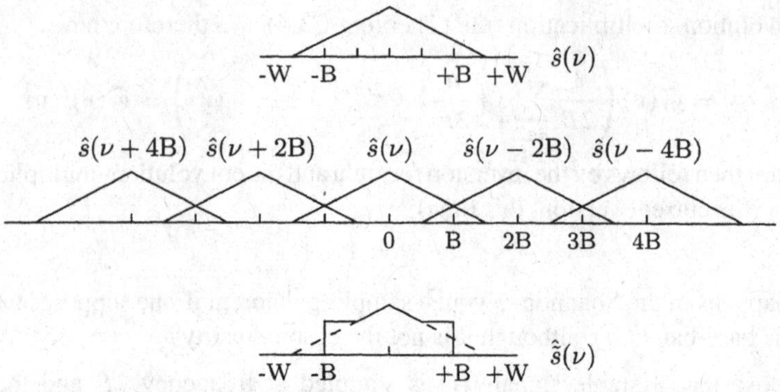

Figure B2.3. Aliasing

EXERCISE **B2.2.** *Let ν_0 and B be such that*

$$0 < 2B < \nu_0,$$

and let $s(t)$ be a stable and continuous base-band (B) signal such that $\sum_{k\in\mathbb{Z}} |s(k/\nu_0)| < \infty$. Consider the train of impulses

$$s_i(t) = \frac{1}{\nu_0} \sum_{n\in\mathbb{Z}} s\left(\frac{n}{\nu_0}\right) \delta\left(t - \frac{n}{\nu_0}\right).$$

Passing this train through a low-pass $(\nu_0 + B)$, one obtains a signal $a(t)$. Passing this train through a low-pass (B), one obtains a signal $b(t)$.

Show that

$$a(t) - b(t) = 2\, s(t) \cos(2\pi \nu_0 t).$$

(We have therefore effected the frequency transposition of the original signal.)

The following exercise gives a version of the sampling theorem for band-pass signals.

EXERCISE **B2.3.** *Let $\nu_0 = 2KB$ for some integer $K \geq 1$, and let $m(t)$ be a stable base-band (B) signal. Consider the frequency-transposed version of this signal, that is, $s(t) = m(t)\cos(2\pi \nu_0 t)$. Suppose that*

$$\sum_{n\in\mathbb{Z}} \left| m\left(\frac{n}{2B}\right)\right| < \infty.$$

Show that if the impulse train

$$\frac{1}{2B} \sum_{n\in\mathbb{Z}} s\left(\frac{n}{2B}\right) \delta\left(t - \frac{n}{2B}\right)$$

is filtered by a low-pass (B), we then recover the original signal $m(t)$.

Oversampling

We have seen the effects of inadapted sampling, that is, sampling at a too slow rate (undersampling) that results in aliasing, or spectrum folding. We now show that oversampling can be exploited to obtain faster rates of convergence in the reconstruction formula.

Assume that the situation of the Shannon–Nyquist theorem prevails; in particular, we have the reconstruction formula (58). The quantity sinc $(2Bt - n)$ therein is of the order of $1/n$ in absolute value and of alternating sign. Therefore, the speed of convergence of the series on the right-hand side is, roughly, comparable to that of

$$\sum_{n\in\mathbb{Z}} \frac{(-1)^n}{n} s\left(\frac{n}{2B}\right).$$

In order to accelerate convergence, one can use oversampling in the following way.

Assume that $\text{supp}(\hat{s}(\nu))$ is contained in the frequency interval $[-W, +W]$ for some $0 < W < \infty$. In formula (57) of Theorem B2.1, choose

$$B = (1+\alpha)W \tag{62}$$

for some $\alpha > 0$ and take any integrable function $T(\nu)$ such that

$$T(\nu) = 1 \quad \text{if } \nu \in [-W, +W]. \tag{63}$$

The resulting signal is then a perfect replica of $s(t)$ since

$$\left\{\sum_{j\in\mathbb{Z}} \hat{s}(\nu + j2B)\right\} T(\nu) = \hat{s}(\nu).$$

Therefore, if we sample at a rate $2B$ larger than the Nyquist rate $2W$, and then filter the resulting train of impulses with a filter of impulse response $h(t)$, we obtain, after division by $2B$, the signal

$$\frac{1}{2B}\sum_{n\in\mathbb{Z}} s\left(\frac{n}{2B}\right) h\left(t - \frac{n}{2B}\right),$$

which is a replica of the original signal $s(t)$ provided the frequency response of the filter verifies condition (63).

EXERCISE **B2.4.** *Suppose that*

$$T(\nu) = \frac{\nu + B}{B - W} 1_{[-B,-W]} + 1_{[-W,+W]} + \frac{-\nu + B}{B - W} 1_{[+B,+W]}.$$

Give the corresponding impulse response, and study the rate of convergence of the series on the right-hand side of (60).

The series in the reconstruction formula can decay faster by choosing a smoother frequency response $T(\nu)$, since increasing the smoothness of a function increases the decay of its Fourier transform.

B2·2 Another Approach to Sampling

This section presents another approach, more direct and with a broader scope, to sampling. It acknowledges the fact that a signal is a combination of complex sinusoids and therefore starts by obtaining the sampling theorem for this type of elementary signals.

Sampling a Single Sinusoid

Consider the signal

$$s(t) = e^{2i\pi\lambda t},$$

where $\lambda \in \mathbb{R}$. This signal is neither stable nor of finite energy, and therefore it does not fit into the framework of the L^1- and L^2-versions of Shannon's sampling theorem. However, the Shannon–Nyquist formula remains essentially true.

THEOREM **B2.4.** *For all $t \in \mathbb{R}$ and all $\lambda \in (-1/2T, +1/2T)$,*

$$e^{2i\pi\lambda t} = \sum_{n\in\mathbb{Z}} e^{2i\pi\lambda nT} \frac{\sin\left(\frac{\pi}{T}(t-nT)\right)}{\frac{\pi}{T}(t-nT)}. \tag{64}$$

For all $B < 1/2T$, the convergence is uniform in $t \in [-B, +B]$.

Proof: We first prove that for all $\lambda \in \mathbb{R}$ and all $t \in (-1/2T, +1/2T)$,

$$e^{2i\pi\lambda t} = \sum_{n\in\mathbb{Z}} e^{2i\pi nTt} \frac{\sin\left(\frac{\pi}{T}(\lambda-nT)\right)}{\frac{\pi}{T}(\lambda-nT)}, \tag{65}$$

where the series converges uniformly for all $t \in [-B, +B]$ for any $B < 1/2T$. The result then follows by exchanging the roles of t and λ.

Let $g(t)$ be the $1/2T$-periodic function equal to $e^{2i\pi\lambda t}$ on $(-1/2T, +1/2T]$. The series in (65) is the Fourier series of $g(t)$. We must therefore show uniform pointwise convergence of this Fourier series to the original function.

Without loss of generality, we do this for the Fourier series of the 2π-periodic function equal to e^{iat} on $(-\pi, +\pi]$, where the convergence is uniform on any interval $[-c, +c] \subset (-\pi, +\pi)$. By (39) of Section A3·1, it suffices to show that

$$\lim_{n\uparrow\infty} \int_{-\pi}^{+\pi} |e^{ia(t-s)} - e^{iat}| \frac{\sin((n+\frac{1}{2})s)}{\sin(s/2)} ds = 0$$

uniformly on $[-c, +c]$. Equivalently,

$$\lim_{n\uparrow\infty} \int_{-\pi}^{+\pi} \frac{|\sin(as/2)|}{\sin(s/2)} \sin((n+\tfrac{1}{2})s)\, ds = 0.$$

This is true, for instance, by the extended Riemann–Lebesgue lemma A1.3. ■

In particular, if $s(t)$ is a trigonometric signal,

$$s(t) = \sum_{k=1}^{M} \gamma_k e^{2i\pi \nu_k t}, \tag{66}$$

where $\gamma_k \in \mathbb{C}$, $\nu_k \in \mathbb{R}$, and if T satisfies

$$\frac{1}{2T} > \sup\{|\nu_k| : 1 \le k \le M\}, \tag{67}$$

we have the Shannon–Nyquist reconstruction formula

$$s(t) = \lim_{N\uparrow\infty} \sum_{-N}^{+N} s(nT) \frac{\sin\left(\frac{\pi}{T}(t-nT)\right)}{\frac{\pi}{T}(t-nT)}. \tag{68}$$

EXERCISE **B2.1.** *With a single sinusoid check that you really need the strict inequality in* (67).

Sampling a Decomposable Signal

The following extension of the sampling theorem for sinusoids is now straightforward:

THEOREM **B2.5.** *Let μ be a nonnegative, finite measure on $[-B, +B]$, where $0 < B < \infty$, and define the signal*

$$s(t) = \int_{[-B,+B]} e^{2i\pi\nu t}\,\mu(d\nu). \tag{69}$$

Then for any $T < 1/2B$ and for all $t \in \mathbb{R}$,

$$s(t) = \sum_{n\in\mathbb{Z}} s(nT) \frac{\sin\left(\frac{\pi}{T}(t-nT)\right)}{\frac{\pi}{T}(t-nT)}. \tag{70}$$

Proof: Since μ is finite and the convergence in (64) is uniform in $\lambda \in [-B, +B]$,

$$s(t) = \int_{[-B,+B]} \left\{ \sum_{n\in\mathbb{Z}} e^{2i\pi\nu nT} \frac{\sin\left(\frac{\pi}{T}(t-nT)\right)}{\frac{\pi}{T}(t-nT)} \right\} \mu(d\nu)$$

$$= \sum_{n\in\mathbb{Z}} \left\{ \int_{[-B,+B]} e^{2i\pi\nu nT}\,\mu(d\nu) \right\} \frac{\sin\left(\frac{\pi}{T}(t-nT)\right)}{\frac{\pi}{T}(t-nT)}. \qquad \blacksquare$$

This closes for the moment our study of the Shannon–Nyquist sampling theory. It will be completed in Section B3·2 by the theorem of equivalence of analog and digital filtering, and in Section C2·2 by the L^2-version of the sampling theorem.

B2·3 Intersymbol Interference

As a further illustration of the weak Poisson sum formula, we consider the problem of intersymbol interference in digital communication. It does not belong to the Shannon–Nyquist sampling theory, however, it does concern sampling.

Pulse Amplitude Modulation and the Nyquist Condition

In a certain type of digital communication system one transmits "discrete" information consisting of a sequence $\{a_n\}_{n\in\mathbb{Z}}$, of real or complex numbers, in the form of an analog signal

$$s(t) = \sum_{n\in\mathbb{Z}} a_n\, g(t - nT), \tag{71}$$

where $g(t)$ is a real or complex function (the "pulse"). Such a "coding" of the information sequence is referred to as *pulse amplitude modulation*. Here, $T > 0$ determines the rate of transmission of information and also the rate at which the information is extracted at the receiver.

EXERCISE **B2.2.** *Assume that $g(t)$ is a stable signal with FT $\hat{g}(\nu)$ and that a_n is stable, with transfer function $A(z)$. Show that $s(t)$ is stable, and give its FT in terms of $\hat{g}(\nu)$ and $A(z)$.*

At time kT the receiver extracts the sample

$$s(kT) = \sum_{n\in\mathbb{Z}} a_n g(kT - nT),$$

that is,

$$a_k g(0) + \sum_{\substack{j\in\mathbb{Z}\\ j\neq 0}} a_{k-j}\, g(jT).$$

If one only wants to obtain a_k from the sample $s(kT)$, the term

$$\sum_{\substack{j\in\mathbb{Z}\\ j\neq 0}} a_{k-j} g(jT)$$

is parasitic. This term disappears for every sequence a_k if and only if

$$g(jT) = 0 \qquad \text{for all } j \neq 0. \tag{72}$$

It turns out that this is equivalent to

$$\sum_{n\in\mathbb{Z}} \hat{g}\left(\nu - \frac{n}{T}\right) = T g(0). \tag{73}$$

The weak version of the Poisson sum formula of Section A2·2 is actually all that is needed to prove the result. Indeed,

THEOREM **B2.6.** *Let $g(t)$ be a continuous and integrable function, and assume that its FT $\hat{g}(\nu)$ is in $L^1_{\mathbb{C}}(\mathbb{R})$. The following two conditions are equivalent:*

(a) $g(jT) = 0$ *for all* $j \in \mathbb{Z}$, $j \neq 0$;

(b) $\sum_{n\in\mathbb{Z}} \hat{g}(\nu + \frac{n}{T}) = \text{const.}$ *almost everywhere.*

Proof: By the weak version of the Poisson sum formula of Section A2·2, $Tg(-nT)$ is the nth Fourier coefficient of $\sum \hat{g}(\nu + n/T)$ (Note that the continuity condition on $g(t)$ is used here.) Therefore, if (b) is true, then (a) is necessarily true. Conversely, if (a) is true, then the sequence $\{Tg(-nT)\}_{n\in\mathbb{Z}}$ is the sequence of Fourier coefficients of two functions, the constant function equal to $Tg(0)$, and $\sum_{n\in\mathbb{Z}} \hat{g}(\nu + n/T)$, and therefore the two functions must be equal almost everywhere. ■

Condition (73) is the *Nyquist condition* for the absence of intersymbol interference.[2]

The pulses $g(t)$ used in communications always, for reasons both technological and operational (bandwidth resources), have a restricted frequency band $[-W, +W]$. The Nyquist condition (73) can be satisfied only if

$$2W \geq \frac{1}{T}. \tag{74}$$

Therefore, if transmission without interference between symbols is required, the minimal bandwidth is

$$2W = 2B \stackrel{\text{def}}{=} \frac{1}{T}.$$

In this case, there is no other choice for the corresponding pulse than

$$\hat{g}(\nu) = \frac{1}{2B} 1_{[-B,+B]}(\nu),$$

that is,

$$g(t) = \frac{\sin(2\pi Bt)}{2\pi Bt}. \tag{75}$$

One disadvantage of such a pulse is linked to questions of numerical stability. Indeed, let us assume that the sampling of $s(t)$ is not carried out at the time kT but at the time $kT + \Delta$, where $\Delta > 0$. We obtain

$$s(kT + \Delta) = a_k \frac{\sin(2\pi B\Delta)}{2\pi B\Delta} + \sum_{j\neq 0} a_{k-j} \frac{\sin(2\pi B\Delta)}{2\pi B(\Delta - jT)}.$$

We see that the error

$$|s(kT + \Delta) - a_k| \tag{76}$$

does not stay bounded for all bounded sequences $\{a_k\}$, because

$$\sum_{j\neq 0} \left| \frac{1}{\Delta - jT} \right| = \infty. \tag{77}$$

[2]See H. Nyquist, (1928), Certain Topics of Telegraph Transmission Theory, *Trans. Amer. Inst. Elec. Eng.*, **47**, 617–644.

A better pulse from this point of view is the "raised cosine"

$$g(t) = \text{sinc}\,(2Bt)\,\frac{\cos(2\pi Bt)}{1-16B^2t^2}, \tag{78}$$

whose FT is

$$\hat{g}(\nu) = \cos^2\left(\frac{\pi\nu}{4B}\right) 1_{[-2B,+2B]}(\nu). \tag{79}$$

In fact,

$$s(kT+\Delta) = a_k \frac{\text{sinc}\,(4B\Delta)}{1-16B^2\Delta^2} + \sum_{j\neq 0} a_{n-j} \frac{\sin(4\pi B\Delta)}{4\pi B(\Delta - jT)(1-16B^2(\Delta - jT)^2)},$$

and the error (76) is seen to remain bounded whatever the bounded sequence $\{a_k\}$.

Partial Response Signaling

Another disadvantage of the pulse (75) is that one cannot realize signals with an FT that has an "infinite slope" (at $-B$ and $+B$).

We shall see that, with clever encoding, we can attain the Nyquist limit (74) (which says that in order to transmit a "symbol" a_n every T seconds without intersymbol interference, a bandwidth of at least $2W = 2B \stackrel{\text{def}}{=} 1/T$ is needed), *without* resorting to an unrealizable pulse (with a very large slope).

For example, in the *duobinary encoding* technique, instead of transmitting (7), one transmits

$$s'(t) = \sum_{n\in\mathbb{Z}} (a_n + a_{n+1})g(t-nT), \tag{80}$$

that is,

$$s'(t) = \sum_{n\in\mathbb{Z}} a_n g'(t-nT), \tag{81}$$

where

$$g'(t) = g(t) + g(t+T). \tag{82}$$

With the pulse (75) of minimal bandwidth $2B$, starting from (80) we obtain

$$s'(kT) = a_k + a_{k-1} = c_k,$$

and from the sequence $\{c_k\}$ and the initial datum a_0 we recover the sequence $\{a_k\}$. The interest of this technique is that we do not seek to implement $s'(t)$ in the form (80) using the unrealizable pulse $g(t)$, but rather in the form (81) with a realizable pulse $g'(t)$. Indeed,

$$\begin{aligned}\widehat{g'}(\nu) &= (1+e^{-2i\pi\nu T})\hat{g}(\nu)\\ &= 2T\cos(\pi\nu T)e^{-2i\pi\nu T} 1_{[-B,+B]}(\nu).\end{aligned}$$

This pulse has minimal bandwidth $2B$, and, furthermore, it is easier to realize, not having an infinite slope.

The above is a particular case of the technique of *partial response signaling*.[3] The general principle is the following: We pretend to use the unrealizable pulse $g(t)$ given by (75), but in (71) we replace the symbol a_n by an encoding c_n, say, a linear encoding

$$c_n = a_n + \gamma_1 a_{n-1} + \ldots + \gamma_{n-k} a_{n-k}\,, \tag{83}$$

which gives

$$s'(t) = \sum_{n \in \mathbb{Z}} c_n g(t - nT).$$

In order to realize $s'(t)$ it is rewritten in the form

$$s'(t) = \sum_{n \in \mathbb{Z}} a_n g'(t - nT),$$

where

$$g'(t) = g(t) + \gamma_1 g(t + T) + \ldots + \gamma_k g(t + kT) \tag{84}$$

is a base-band (B) pulse, in general easily realizable, with FT

$$\hat{g}'(\nu) = T(\nu)\hat{g}(\nu), \tag{85}$$

where

$$T(\nu) = 1 + \sum_{j=1}^{k} \gamma_j e^{-2i\pi\nu jT} = P(e^{-2i\pi\nu T}) \tag{86}$$

and

$$P(z) = 1 + \sum_{j=1}^{k} \gamma_j z^j. \tag{87}$$

By sampling at the time $t = kT$ we obtain

$$s'(kT) = P(z) a_k = c_k.$$

We shall see in Section B3·2 that the sequence $\{a_k\}$ is deduced from the sequence $\{c_k\}$ by inverse filtering

$$a_k = \frac{1}{P(z)} c_k \tag{88}$$

(we assume that $1/P(z)$ is stable and therefore that the corresponding filter is causal; these notions are discussed in detail in Section B3·2).

---

[3]See A. Lender (1981), Correlative (Partial Response) Techniques and Applications to Radio Systems, in Feher, K. (ed.), *Digital Communications: Microwave Applications (Prentice–Hall: Englewood Cliffs, NJ), Ch. 7.*.

B2·4 The Dirac Formalism

Do We Need Distributions Theory Here?

In the applied literature, the Dirac formalism of generalized functions is used profusely. It consists of a small set of symbolic rules that are justified by the classical Fourier theory of the previous chapters. In signal processing, the Dirac formalism culminates in the formula giving the FT of a Dirac comb. We shall see that the Poisson sum formula is, for all practical purposes, all that is needed to deal with such a mathematical object in a rigorous way.

We shall see in the next part that the Fourier theory has in the Hilbert space framework a high degree of formal beauty. There was yet another important step to be made in this direction. The physicists had introduced a very useful tool, the Dirac generalized function, associated with a formal calculus that was rather pleasant to use, but that lacked mathematical foundations. These were established by Laurent Schwartz, with the elegant theory of distributions (or generalized functions) and the equally elegant Fourier theory of tempered distributions.[4]

Most engineers are familiar with the so-called Dirac function $\delta(t)$, which is "defined" by the property

$$\int_{\mathbb{R}} \varphi(t)\delta(t)\,\mathrm{d}t = \varphi(0),$$

for all functions $\varphi(t)$. They are aware that there exists no such function in the usual sense with such property, and they take the above formula as a symbolic way of dealing with a limit situation. In the "prelimit," $\delta(t)$ is replaced by a proper function, depending on a parameter, say, n. There are many choices for this proper function $\delta_n(t)$, the simplest one being

$$\delta_n(t) = n\,\mathrm{rec}_{\frac{1}{n}}(t).$$

Then for sufficiently regular function $\varphi(t)$ (say, continuous),

$$\lim_{n\uparrow\infty} \int_{\mathbb{R}} \varphi(t)\delta_n(t)\,\mathrm{d}t = \varphi(0).$$

Thus, in this point of view, the Dirac function is the limit of proper functions becoming more and more concentrated around the origin of times while their integral remains equal to 1. Another candidate with these properties is the Gaussian pulse that we have already encountered in the proof of the inverse Fourier formula:

$$h_\sigma(t) = \frac{1}{\sigma\sqrt{2\pi}}\, e^{-\frac{t^2}{2\sigma^2}},$$

where this time the positive parameter σ tends to zero. Observe that the FT of both $\delta_n(t)$ and $h_\sigma(t)$ (which we have previously computed) converge pointwise, as the

[4] *Théorie des Distributions*, Vols. 1 and 2, 1950–1, Hermann, Paris.

corresponding parameters tend to the appropriate limits, to 1. This is consistent with the formal computation of the FT of the Dirac function

$$\hat{\delta}(\nu) = \int_{\mathbb{R}} \delta(t) e^{-2i\pi\nu t}\, dt = e^{-2i\pi\nu 0} = 1.$$

Another generalized function that is omnipresent in the signal processing literature is the Dirac comb (indeed a cosmetic tool!), also called the Dirac pulse train. It is the T-periodic generalized function

$$\Delta_T(t) = \sum_{n\in\mathbb{Z}} \delta(t - nT).$$

If we formally compute its nth Fourier coefficient, we obtain

$$\frac{1}{T}\int_0^T \delta(t) e^{-2i\pi\frac{n}{T}t}\, dt = 1.$$

Now if we write the corresponding formal Fourier series

$$\frac{1}{T}\sum_{n\in\mathbb{Z}} e^{2i\pi\frac{n}{T}\nu},$$

we observe that its convergence is rather problematic. We can, however, pursue the heuristics, and consider that the latter sum is the limit as $N \to \infty$ of the $1/T$-periodic function

$$\frac{1}{T}\sum_{-N}^{+N} e^{2i\pi\frac{n}{T}\nu},$$

which is the Dirichlet kernel

$$\frac{1}{T}\frac{\sin(2\pi(N + \frac{1}{2}T)}{\sin(\pi T)}.$$

Graphically, up to a multiplicative factor $1/T$, such a function looks in the vicinity of 0 like a Dirac function: As $N \to \infty$, it becomes more and more concentrated around 0, and its integral in a neighborhood of 0 tends to 1. Therefore, at the limit we have, invoking the $1/T$-periodicity, the Fourier transform of the Dirac comb

$$\hat{\Delta_T}(\nu) = \frac{1}{T}\sum_{n\in\mathbb{Z}} \delta\left(\nu - \frac{n}{T}\right).$$

This overdose of heuristics may well be fatal for the more critical mind. However, in most basic courses in signal analysis, it is administered with the best intentions, with the excuse that it saves the student from a painful exposition to distributions theory. This apology of mathematical euthanasy is founded on wrong premices. The first question that one should ask is: Do we need the Dirac comb in signal analysis? Looking back at the previous chapters, we can immediately answer NO. It is not needed to derive the Shannon–Nyquist theorem, because the Poisson formula is all that is needed there. Is the Poisson formula harder than distributions theory? Again, the answer is NO, without surprise, because the distributions theory version of the Poisson sum formula is only a small chapter of distributions theory. (I shall add that the heuristic derivation of the Poisson sum formula—see the comment following the statement of Theorem A2.3 of Chapter 1—is much more convincing than the usual heuristic derivation of the FT of the Dirac comb.)

In fact, the reader may skip this chapter and proceed to Chapters 3 and 4 without damage. On the other hand, the Fourier transform of the Dirac comb is part of a well-established tradition in signal analysis that is bound to be eternal due to its aesthetic appeal. I have therefore devoted the next section to the expression of the classical results of Fourier analysis in the Dirac formalism. It is, however, a purely symbolic analysis.

The Dirac Generalized Function

The principal formal object of the Dirac formalism is the Dirac generalized function $\delta(t)$, and the first formal rule is the symbolic formula

$$\int_{\mathbb{R}} \varphi(t)\delta(t-a)\,\mathrm{d}t = \varphi(a). \tag{D1}$$

EXAMPLE **B2.1.** *By the first symbolic rule,*

$$\int_{\mathbb{R}} e^{-2i\pi\nu t}\delta(t-a)\,\mathrm{d}t = e^{-2i\pi\nu a},$$

that is,

$$\delta(t-a) \overset{\mathrm{FT}}{\to} e^{-2i\pi\nu a}.$$

In particular, the Fourier transform of the Dirac generalized function is the constant function equal to 1.

EXAMPLE **B2.2.** *Let $x(t)$ be a T-periodic signal, and let $\{\hat{x}_n\}$, $n \in \mathbb{Z}$, be the sequence of its Fourier coefficients. In Section A2·1 we defined the FT $x(t)$ symbolically, by*

$$\hat{x}(\nu) = \sum_{n\in\mathbb{Z}} \hat{x}_n \delta\left(\nu - \frac{n}{T}\right).$$

Using the symbolic formula

$$x(t) = \int_{\mathbb{R}} e^{2i\pi\nu t}\hat{x}(\nu)\,\mathrm{d}\nu$$

and the symbolic rule (D1), *we then have*

$$\begin{aligned} x(t) &= \sum_{n\in\mathbb{Z}} \int_{\mathbb{R}} e^{2i\pi\nu t}\hat{x}_n\delta\left(\nu - \frac{n}{T}\right)\mathrm{d}\nu \\ &= \sum_{n\in\mathbb{Z}} \hat{x}_n e^{2i\pi\frac{n}{T}t}, \end{aligned}$$

and we recover the inversion formula of Section A2·1.

EXAMPLE **B2.3.** *Let $x(t)$ be as in the previous example. If it is the input of a filter with (stable) impulse response $h(t)$ and with frequency response $T(\nu) = \hat{h}(\nu)$, symbolic calculations give for the output*

$$y(t) = \int_{\mathbb{R}} \hat{h}(\nu)\hat{x}(\nu)e^{2i\pi\nu t}\,d\nu$$

$$= \sum_{n\in\mathbb{Z}} \hat{x}_n \int_{\mathbb{R}} \hat{h}(\nu)e^{2i\pi\nu t}\delta\left(\nu - \frac{n}{T}\right)d\nu,$$

that is,

$$y(t) = \sum_{n\in\mathbb{Z}} \hat{h}\left(\frac{n}{T}\right)\hat{x}_n e^{2i\pi\frac{n}{T}t}.$$

The sequence $\{\hat{y}_n\}$ of Fourier coefficients of $y(t)$ is thus

$$\hat{y}_n = \hat{h}\left(\frac{n}{T}\right)\hat{x}_n,$$

a result that we already know.

The FT of the Dirac Comb

Consider the Dirac comb

$$\Delta_T(t) = \sum_{n\in\mathbb{Z}} \delta(t - nT).$$

The second symbolic fomula, that we now introduce, gives the FT of this generalized function:

$$\Delta_T(t) \xrightarrow{\text{FT}} \frac{1}{T}\Delta_{\frac{1}{T}}(\nu). \tag{D2}$$

EXAMPLE **B2.4.** *The Poisson sum formula. The formal Plancherel–Parseval equality*

$$\int_{\mathbb{R}} \varphi(t)\Delta_T(t)\,dt = \int_{\mathbb{R}} \widehat{\varphi}(\nu)\widehat{\Delta_T}(\nu)\,d\nu$$

gives, upon substituting into it

$$\widehat{\Delta_T}(\nu) = \frac{1}{T}\sum_{n\in\mathbb{Z}} \delta\left(\nu - \frac{n}{T}\right),$$

the Poisson sum formula

$$\sum_{n\in\mathbb{Z}} \varphi(nT) = \frac{1}{T}\sum \widehat{\varphi}\left(\frac{n}{T}\right).$$

Multiplication Rule

The third symbolic formula of the Dirac formalism concerns the multiplication of a Dirac generalized function by a function in the usual sense:

$$s(t)\delta(t - a) \equiv s(a)\delta(t - a). \tag{D3}$$

This rule is consistent with the first rule, in that

$$\int_{\mathbb{R}} s(t)\delta(t-a)\varphi(t)\,dt = s(a)\varphi(a) = \int_{\mathbb{R}} s(a)\delta(t-a)\varphi(t)\,dt\,.$$

EXAMPLE **B2.5.** *Sampling and Spectrum Folding. The train of sampled pulses*

$$s_i(t) = \sum_{n\in\mathbb{Z}} s(nT)\delta(t-nT)$$

may, in view of (D3), *be formally written*

$$s_i(t) = s(t)\Delta_T(t).$$

Its symbolic FT is therefore

$$\begin{aligned}\hat{s}_i(\nu) &= \hat{s}(\nu) * \widehat{\Delta_T}(\nu)\\ &= \int_{\mathbb{R}} \hat{s}(\nu-\mu)\left(\frac{1}{T}\sum_{n\in\mathbb{Z}}\delta\left(\mu-\frac{n}{T}\right)\right)d\mu,\end{aligned}$$

that is,

$$\hat{s}_i(\nu) = \frac{1}{T}\sum_{n\in\mathbb{Z}}\hat{s}\left(\nu-\frac{n}{T}\right).$$

If we input $T s_i(t)$ *into a low-pass* $[-1/T, +1/T]$ *filter, we therefore obtain at the output the signal* $\tilde{s}(t)$ *with FT*

$$\hat{\tilde{s}}(\nu) = \sum_{n\in\mathbb{Z}}\hat{s}\left(\nu-\frac{n}{T}\right)1_{[-\frac{1}{T},+\frac{1}{T}]}(\nu).$$

This is the equation describing spectrum folding (see Theorem B2.3).

EXAMPLE **B2.6.** *The FT of a Radar Return Signal. Let us consider the signal*

$$\begin{aligned}s(t) &= \left(\sum_{n\in\mathbb{Z}} h(t-nT)\right) f(t)\\ &= (h(t) * \Delta_T(t))f(t) = v(t)f(t).\end{aligned}$$

Its FT is (by the convolution–multiplication formula)

$$\hat{s}(\nu) = \int_{\mathbb{R}} \hat{v}(\mu)\hat{f}(\nu-\mu)\,d\mu.$$

Now, on using the rule (D3)*:*

$$\begin{aligned}\hat{v}(\nu) &= \hat{h}(\nu)\widehat{\Delta}_T(\nu)\\ &= \frac{1}{T}\sum_{n\in\mathbb{Z}}\hat{h}(\nu)\delta\left(\nu-\frac{n}{T}\right)\\ &= \frac{1}{T}\sum_{n\in\mathbb{Z}}\hat{h}\left(\frac{n}{T}\right)\delta\left(\nu-\frac{n}{T}\right).\end{aligned}$$

On the other hand,

$$\int_{\mathbb{R}} \delta\left(\mu - \frac{n}{T}\right) \hat{f}(\nu - \mu)\,\mathrm{d}\mu = \hat{f}\left(\nu - \frac{n}{T}\right).$$

Thus we have

$$\hat{s}(\nu) = \frac{1}{T} \sum_{n \in \mathbb{Z}} \hat{h}\left(\frac{n}{T}\right) \hat{f}\left(\nu - \frac{n}{T}\right),$$

that is, Eq. (71) *of Section B2.3.* ■

The examples above show how the Dirac symbolic calculus formally accounts for calculations of Fourier transforms. This symbolic calculus retrieves formulas already proven in the framework of the classical Fourier theory in L^1, formulas that have been proved under certain conditions of regularity, and of integrability or summability. The symbolic calculus does not say under what conditions the final symbolic formulas have a meaning, nor in what sense they must be interpreted (equalities almost everywhere? in L^1?). For this reason, the Dirac symbolic calculus must be used with precaution. From a mnemonic point of view, it can be useful, as it allows one to obtain some formulas very quickly, and "generally" these formulas are correct under conditions that are "almost always" satisfied in practice.

However, let us emphasize once more the fact that these formulas have been obtained rigorously within the framework of Fourier transforms in L^1.

B3

Digital Signal Processing

B3·1 The DFT and the FFT Algorithm

The DFT

Suppose we need to compute numerically the FT of a stable signal $s(t)$. In practice only a finite vector of samples is available,

$$s = (s_0, \ldots, s_{N-1}),$$

where $s_n = s(n\Delta)$. The Fourier sum of this vector evaluated at pulsations $\omega_k = 2k\pi/N$ is the discrete Fourier transform (DFT).

DEFINITION **B3.1.** *The DFT of* $s = (s_0, \ldots, s_{N-1})$ *is the vector* $S = (S_0, \ldots, S_{N-1})$, *where*

$$S_k = \sum_{n=0}^{N-1} s_n e^{-i(2\pi kn/N)}.$$

The DFT is an approximation of the FT, the quality of which depends on the parameters N and Δ. The first question to ask is: How to choose these parameters to attain a given precision? As we shall see, the answer is given by the Poisson sum formula. For the time being, we shall give the basic properties of the DFT without reference to a sampled signal.

Let $a = (a_0, \ldots, a_{N-1})$ be a finite sequence of complex numbers. For the Nth root of unity, we adopt the following notation:

$$w_N = e^{-i\frac{2\pi}{N}}.$$

The finite sequence $A = (A_0, \dots, A_{N-1})$ defined by

$$A_m = \sum_{n=0}^{N-1} a_n w_N^{mn} \tag{89}$$

is the DFT of $a = (a_0, \dots, a_{N-1})$.

THEOREM **B3.1.** *We have the inversion formula*

$$a_n = \frac{1}{N} \sum_{m=0}^{N-1} A_m w_N^{-mn}. \tag{90}$$

Proof:

$$\sum_{m=0}^{N-1} A_m w_N^{-mn} = \sum_{m=0}^{N-1} w_N^{-mn} \left(\sum_{k=0}^{N-1} a_k w_N^{mk} \right)$$

$$= \sum_{k=0}^{N-1} a_k \sum_{m=0}^{N-1} w_N^{m(k-n)}.$$

But if $k \neq n$,

$$\sum_{m=0}^{N-1} w_N^{m(k-n)} = \frac{w_N^{N(k-n)} - 1}{w_N^{k-n} - 1} = 0$$

since $w_N^{Nr} = 1$ when $r \neq 0$; on the other hand, for $k = n$,

$$\sum_{m=0}^{N-1} w_N^{m(k-n)} = \sum_{m=0}^{N-1} 1 = N. \qquad \blacksquare$$

If we consider the periodic extensions of the finite sequences $a = (a_0, \dots, a_{N-1})$ and $A = (A_0, \dots, A_{N-1})$, defined by

$$a_{n+kN} = a_n, \qquad A_{m+kN} = A_m, \tag{91}$$

Eqs. (89) and (90) remain valid since $w_N^{(m+kN)n} = w_N^{mn}$.

The sequences a_n and A_m being N-periodic, the domains of the sums (89) and (90) can be shifted arbitrarily. In particular, with $N = 2M + 1$,

$$A_m = \sum_{n=-M}^{+M} a_n w_N^{mn}, \qquad a_n = \frac{1}{2M+1} \sum_{m=-M}^{+M} A_m w_N^{-mn}. \tag{92}$$

In the sequel, we use the above periodic extensions. The relation between the sequences $\{a_n\}$ and $\{A_m\}$ will be symbolized by

$$a_n \underset{N}{\leftrightarrow} A_m. \tag{93}$$

We observe that

$$a^*_{-n} \underset{N}{\leftrightarrow} A^*_m. \tag{94}$$

THEOREM **B3.2.** *If* $a_n \underset{N}{\leftrightarrow} A_m$ *and* $b_n \underset{N}{\leftrightarrow} B_m$, *we have the convolution–multiplication rule*

$$\sum_{k=0}^{N-1} a_k b_{n-k} \underset{N}{\leftrightarrow} A_m B_m. \tag{95}$$

Proof: The proof of (95) consists of a simple verification. In fact,

$$\sum_{n=0}^{N-1}\left(\sum_{k=0}^{N-1} a_k b_{n-k}\right) w_N^{mn} = \sum_{k=0}^{N-1} a_k \left(\sum_{n=0}^{N-1} b_{n-k} w_N^{mn}\right).$$

The change of variable $n - k = r$ gives

$$\sum_{n=0}^{N-1} b_{n-k} w_N^{mn} = w_N^{mk} \sum_{r=-k}^{N-1-k} b_r w_n^{mr}.$$

But because of the N-periodicity of the sequences $\{b_r\}$ and $\{w_N^{mr}\}$, the last sum can be taken from 0 to $N - 1$. Therefore,

$$\sum_{n=0}^{N-1}\left(\sum_{k=0}^{N-1} a_k b_{n-k}\right) w_N^{mn} = \sum_{k=0}^{N-1} a_k \left(w_N^{mk} \sum_{r=0}^{N-1} b_r w_n^{mr}\right) = A_m B_m. \qquad \blacksquare$$

Equation (95) and the inversion formula (90) give

$$\sum_{k=0}^{N-1} a_k b_{n-k} = \frac{1}{N} \sum_{m=0}^{N-1} A_m B_m w_N^{-mn}. \tag{96}$$

Making $n = 0$ in (96) and taking (94) into account, we obtain the Plancherel–Parseval equation for the DFT

$$\sum_{k=0}^{N-1} a_k b_k^* = \frac{1}{N} \sum_{m=0}^{N-1} A_m B_m^*. \tag{97}$$

With $a_n \equiv b_n$ we obtain the energy conservation formula

$$\sum_{k=0}^{N-1} |a_k^2| = \frac{1}{N} \sum_{m=0}^{N-1} |A_m|^2. \tag{98}$$

The Fast Fourier Transform Algorithm

The calculation of the DFT of the sequence $\{a_n\}$ by formula

$$A_m = a_0 + a_1 w_N^m + a_2 w_N^{2m} + \ldots + a_{N-1} w_N^{m(N-1)}$$

requires $N - 1$ multiplications for $m \geq 1$ (none for $m = 0$). If we consider that the cost of an addition is negligible compared with that of a multiplication, the calculation of $A = (A_0, \ldots, A_{N-1})$ thus requires $(N - 1)^2$ computational units, where one unit corresponds to one multiplication. The fast Fourier algorithm,[5], also called the fast Fourier transform (FFT), considerably reduces the computational complexity. It is based on the following remark.

Let $a_n \underset{2N}{\leftrightarrow} A_m$ be a DFT pair (note that we are considering a DFT of order $2N$ with $2N$ terms a_n and $2N$ terms A_m). Define

$$b_n = a_{2n}, \quad c_n = a_{2n+1} \qquad (0 \le n \le N-1),$$

and

$$b_n \underset{N}{\leftrightarrow} B_m, \quad c_n \underset{N}{\leftrightarrow} C_m$$

(the latter DFTs are of order N). A direct calculation shows that

$$A_m = B_m + w_{2N}^m C_m \qquad (0 \le m \le 2N-1). \tag{99}$$

Observing that $B_{m+N} = B_m$, $C_{m+N} = C_m$, and $w_{2N}^{m+N} = -w_{2N}^m$, we can split Eq. (99) in two parts:

$$A_m = B_m + w_{2N}^m C_m \qquad (0 \le m \le N-1) \tag{100}$$

and

$$A_{m+N} = B_m - w_{2N}^m C_m \qquad (0 \le m \le N-1). \tag{101}$$

In order to calculate B_m and C_m for $0 \le m \le N-1$, we need $2(N-1)^2$ computational units. When (100) is used we need $N-1$ additional multiplications. The multiplications in (101) are for free since they were done in (100). In total, the method requires

$$2(N-1)^2 + N - 1 = (N-1)(2N-3)$$

units instead of $(2N-1)^2$ for the direct method. If we have to calculate a DFT of order N such that

$$N = 2^s, \tag{102}$$

the FFT will take $F(N) \le \frac{1}{2}N \log_2 N$ computational units. The result is obtained by induction. Indeed, $F(2) = 1$, and the considerations above show that

$$F(2N) = 2F(N) + N - 1 \le 2F(N) + N.$$

But if $F(N) \le \frac{1}{2}N \log_2 N$, then $2F(N) + N \le N(\log_2 N + 1) = \frac{1}{2}2N \log_2 2N$.

The gain in computational complexity with respect to the direct method is thus of the order of

$$\frac{1}{2}\frac{\log_2 N}{N}.$$

[5] Cooley, J.W., Lewis, P.A.W., and Welch, P.D., The Fast Fourier Transform Algorithm, considerations in the calculation of sine, cosine, and Laplace transforms, *J. Sound Vibrations*, 1970, 12(3), 315–337.

The above discussion just gives the basic idea of the FFT. For a detailed account of the algorithmic aspects of the discrete-time Fourier transform, see, for instance, [B8]. We now turn to the numerical issues behind the DFT.

Numerical Analysis of the DFT

The Poisson sum formula is useful in numerical analysis when approximating a Fourier integral by a Darboux sum, and this is of course related to the finite Fourier transform.

Let us recall the Poisson sum formula, assuming that the conditions of validity are satisfied:

$$\sum_{n\in\mathbb{Z}} s(n\Delta)e^{-2i\pi n\Delta\nu} = \frac{1}{\Delta}\sum_{n\in\mathbb{Z}} \hat{s}\left(\nu + \frac{n}{\Delta}\right). \tag{103}$$

The expression (103) elucidates the relation between the FT $\hat{s}(\nu)$ of the signal $s(t)$ and the DFT of its sampled and truncated version $(s(-M\Delta), \dots, s(+M\Delta))$,

$$\sum_{n=-M}^{+M} s(n\Delta)e^{-2i\pi n\frac{k}{2M+1}}.$$

In fact, letting $\nu = k/[(2M+1)\Delta]$ in (103),

$$\sum_{n\in\mathbb{Z}} s(n\Delta)e^{-i2\pi\frac{nk}{2M+1}} = \frac{1}{\Delta}\sum_{n\in\mathbb{Z}} \hat{s}\left(\frac{n}{\Delta} + \frac{k}{(2M+1)\Delta}\right). \tag{104}$$

THEOREM **B3.3.** *Let $s(t)$ be a signal with support contained in $[-M\Delta, +M\Delta]$. We then have*

$$\sum_{n=-M}^{+M} s(n\Delta)e^{-2i\pi\frac{kn}{2M+1}} = \frac{1}{\Delta}\sum_{n\in\mathbb{Z}} \hat{s}\left(\frac{n}{\Delta} + \frac{k}{(2M+1)\Delta}\right). \tag{105}$$

Proof: Just apply formula (104). ■

If the terms corresponding to the indices $n \neq 0$ in the right-hand side of (105) were null, only the central term

$$\frac{1}{\Delta}\hat{s}\left(\frac{k}{(2M+1)\Delta}\right)$$

would remain. The DFT of $(s(-M\Delta), \dots, s(+M\Delta))$ would then be a sampled version of the FT, that is,

$$\left(\frac{1}{\Delta}\hat{s}(-M\nu_1), \dots, \frac{1}{\Delta}\hat{s}(+M\nu_1)\right),$$

where $\nu_1 = 1/[(2M+1)\Delta]$. But one cannot have a signal $s(t)$ with bounded support which has FT $\hat{s}(\nu)$ also with bounded support. There will thus always be an error, equal to

$$\sum_{\substack{n\in\mathbb{Z}\\ n\neq 0}} \hat{s}\left(\frac{n}{\Delta} + \frac{k}{(2M+1)\Delta}\right).$$

This error is the *aliasing error*. It can be controlled by choosing Δ small enough for $\hat{s}(\nu)$ to be negligible outside the interval $[-B, +B] = [-1/2\Delta, 1/2\Delta]$. But then M must be adjusted so that $s(t)$ remains zero outside $[-M\Delta, +M\Delta]$. Increasing M increases the computational complexity.

We shall retain the approximate relation linking the effective bandwidth $2B = 1/\Delta$, the effective temporal extension $T = 2M\Delta$, and the complexity $N = 2M+1$:

$$2BT \simeq N. \tag{106}$$

B and T are chosen such that $s(t)$ is negligible outside $[-T/2, +T/2]$ and $\hat{s}(\nu)$ is negligible outside $[-B, +B]$. Precision requires large T and large B, in order to capture a large amount of the time–frequency content of the signal. This results in large complexity (measured by N) of the DFT. This in turn requires sophisticated algorithms such as the FFT in order to reduce the computational load.

B3·2 The Z-Transform

Discrete-Time Fourier Transform

A discrete-time signal is, in signal processing, a *sampled signal*. This section gives the basic tools of digital signal processing: the Fourier transform (reducing to a Fourier sum) and the z-transform.

DEFINITION **B3.1.** *A stable discrete-time signal is a sequence $\{x_n\}_{n\in\mathbb{Z}}$ of complex numbers such that*

$$\sum_{n\in\mathbb{Z}} |x_n| < \infty. \tag{107}$$

Its Fourier sum *is the function*

$$\tilde{x}(\omega) = \sum_{k\in\mathbb{Z}} x_k e^{-ik\omega}. \tag{108}$$

We observe that it is a 2π-periodic function. Also, with the same arguments as for the FT of a stable signal (see Section A1·1), we observe that it is continuous and bounded by $\sum_{n\in\mathbb{Z}} |x_n|$.

An inversion formula is available:

THEOREM **B3.1.** *x_n is the nth Fourier coefficient of $\tilde{x}(\omega)$:*

$$x_n = \frac{1}{2\pi}\int_{-\pi}^{+\pi} \tilde{x}(\omega)e^{in\omega}\,d\omega. \tag{109}$$

Proof: Multiply (108) by $e^{in\omega}$ and integrate from $-\pi$ to $+\pi$. ■

EXERCISE **B3.1.** *Let $x_n = s(n/2B)$, where $s(t)$ is a continuous base-band (B) signal with FT $\hat{s}(\nu)$. Show that the Fourier sum associated with $\{x_n\}$ is*

$$\tilde{x}(\omega) = 2B\hat{s}\left(\omega\frac{2B}{2\pi}\right). \tag{110}$$

EXERCISE **B3.2.** *Give the impulse response of the filter with frequency response*

$$\exp(\cos(\omega))e^{i\sin(\omega)}.$$

DEFINITION **B3.2.** *The operation that associates to a stable discrete-time signal x_n the discrete-time signal*

$$y_n = \sum_{k\in\mathbb{Z}} x_k h_{n-k}, \tag{111}$$

where h_n is a stable signal, is called convolutional filtering. *The signal y_n is the output of the convolutional filter $\mathcal{F}$ with* impulse response h_n, *and x_n is the input.*

When the input signal is the unit impulse at 0,

$$\delta_n = \begin{cases} 1 & \text{if } n = 0, \\ 0 & \text{otherwise,} \end{cases} \tag{112}$$

the output is $y_n = h_n$, whence the terminology.

When x_n and h_n are stable, the right-hand side of (111) has a meaning. In fact,

$$\sum_{n\in\mathbb{Z}}\sum_{n\in\mathbb{Z}} |x_k|\,|h_{n-k}| = \sum_{n\in\mathbb{Z}} \left\{ |x_k| \sum_{n\in\mathbb{Z}} |h_{n-k}| \right\}$$
$$= \sum_{k\in\mathbb{Z}} |x_k| \sum_{\ell\in\mathbb{Z}} |h_\ell| < \infty,$$

and, in particular,

$$\sum_{k\in\mathbb{Z}} |x_k|\,|h_{n-k}| < \infty \quad \text{for all } n \in \mathbb{Z}.$$

This also shows that y_n is stable.

DEFINITION **B3.3.** *A* causal, *or* physically realizable, *filter is one such that*

$$h_n = 0 \qquad \text{for all } n < 0. \tag{113}$$

The filter is called causal because if the input x_n is zero for $n \le n_0$ the output y_n is zero for $n \le n_0$. The input–output relation (111) takes, for a causal filter, the form

$$y_n = \sum_{k=-\infty}^{n} x_k h_{n-k}. \tag{114}$$

DEFINITION **B3.4.** *The Fourier sum*

$$\tilde{h}(\omega) = \sum_{n\in\mathbb{Z}} h_n e^{-in\omega} \tag{115}$$

is the frequency response *of the convolutional filter with stable impulse response h_n.*

If we write $\tilde{x}(\omega)$ and $\tilde{y}(\omega)$, respectively, for the Fourier sums of the input x_n and the output y_n, the input–output relation (6) reads

$$\tilde{y}(\omega) = \tilde{h}(\omega)\tilde{x}(\omega). \tag{116}$$

Indeed,

$$\begin{aligned}\tilde{y}(\omega) &= \sum_{n\in\mathbb{Z}} y_n e^{-in\omega}\\ &= \sum_{n\in\mathbb{Z}}\sum_{k\in\mathbb{Z}} x_k h_{n-k} e^{-in\omega}\\ &= \sum_{k\in\mathbb{Z}} \left\{ x_k e^{-ik\omega} \sum_{n\in\mathbb{Z}} h_{n-k} e^{-i(n-k)\omega} \right\}\\ &= \left(\sum_{k\in\mathbb{Z}} x_k e^{-ik\omega}\right)\left(\sum_{\ell\in\mathbb{Z}} h_\ell e^{-i\ell\omega}\right).\end{aligned}$$

■

EXAMPLE **B3.1** (The pure delay). *The input–output relation $x_n \to y_n$ defined by*

$$y_n = x_{n-k}$$

is a homogeneous filtering with impulse response

$$h_n = \begin{cases} 1 & \text{if } n = k,\\ 0 & \text{if } n \neq k,\end{cases}$$

and frequency response

$$\tilde{h}(\omega) = e^{-ik\omega}.$$

EXERCISE **B3.3** (The smoothing filter). *This is the filter defined by the input–output relation*

$$y_n = \frac{1}{2N+1}\sum_{k=-N}^{+N} x_{n-k}.$$

Show that its frequency response is

$$\tilde{h}_N(\omega) = \frac{1}{2N+1}\,\frac{\sin\{(N+\frac{1}{2})\omega\}}{\sin\{\omega/2\}},$$

where $\tilde{h}(0) = 1$. What is $\lim_{N\uparrow\infty}\tilde{h}_N(\omega)$?

Equivalence of Analog and Digital Filtering

It is important to understand how the operation of analog filtering followed by sampling can be performed if one chooses first to sample and then to operate in the sampled (digital) domain. The precise statement and the precise answer are contained in the theorem below.

Let $s(t)$ be a stable continuous signal, base-band (B), sampled at the Nyquist frequency $2B$. We obtain the sampled signal

$$s\left(\frac{n}{2B}\right) = \int_{-B}^{+B} e^{2i\pi\nu\frac{n}{2B}}\hat{s}(\nu)\,d\nu$$
$$= \frac{1}{2\pi}\int_{-\pi}^{+\pi} e^{i\pi\omega}2B\hat{s}\left(\frac{B}{\pi}\omega\right)d\omega$$

(the inversion formula can be applied because $\hat{s}(\nu)$ is integrable, having a bounded support; on the other hand, the equality of $s(t)$ and $\int_{\mathbb{R}} \hat{s}(\nu)e^{2i\pi\nu t}\,dt$ holds for all t, since both quantities are continuous). It is further assumed that

$$\sum_{n\in\mathbb{Z}} |s\left(\frac{n}{2B}\right)| < \infty, \tag{117}$$

and therefore, the Fourier sum of $s(n/2B)$ is $2B\hat{s}((B/\pi)\omega)$.

THEOREM **B3.2.** *Let $x(t)$ and $h(t)$ be stable continuous signals, base-band (B), both satisfying the condition of type* (117). *Then*

$$y\left(\frac{n}{2B}\right) = \frac{1}{2B}\sum_{n\in\mathbb{Z}} h\left(\frac{k}{2B}\right)x\left(\frac{k}{2B}\right). \tag{118}$$

This is the theorem of the equivalence of analog and digital filtering.

Proof: The discrete-time signal

$$y_n = \frac{1}{2B}\sum_{k\in\mathbb{Z}} h\left(\frac{k}{2B}\right)x\left(\frac{n-k}{2B}\right)$$

is stable, and its Fourier sum is

$$\tilde{y}(\omega) = 2B\hat{h}\left(\frac{B}{\pi}\omega\right)\hat{x}\left(\frac{B}{\pi}\omega\right).$$

Hence we have

$$y_n = \frac{1}{2\pi}\int_{-\pi}^{+\pi} 2B\hat{h}\left(\frac{B}{\pi}\omega\right)\hat{x}\left(\frac{B}{\pi}\omega\right)e^{in\omega}\,d\omega$$
$$= \int_{-B}^{+B} \hat{h}(\nu)\hat{x}(\nu)e^{2i\pi\nu\frac{n}{2B}}\,d\nu.$$

On the other hand, for the analog signal

$$y(t) = \int_{\mathbb{R}} h(t-s)x(s)\,ds, \tag{119}$$

we have the inversion formula

$$y(t) = \int_{-B}^{+B} e^{2i\pi\nu t}\hat{x}(\nu)\hat{h}(\nu)\,d\nu,$$

and therefore $y_n = y(n/2B)$. ■

Transfer Functions

To every discrete-time signal x_n is associated its *formal z-transform*, which is the formal series

$$X(z) = \sum_{n\in\mathbb{Z}} x_n z^n. \tag{120}$$

The formal z-transform of the impulse response h_n of a convolutional filter is the *formal transfer function* of the filter considered:

$$H(z) = \sum_{n\in\mathbb{Z}} h_n z^n. \tag{121}$$

The input–output relation (111) reads as a function of the z-transforms of x_n, y_n, and h_n

$$Y(z) = H(z)X(z). \tag{122}$$

Note, however, that the z-transform of a signal only takes a meaning as a function of $z \in \mathbb{C}$ if one gives the domain of convergence of the series defining it.

We use the *unit delay* operation z defined symbolically by

$$z^k x_n = x_{n-k}\,.$$

With this notation the relation (6) is written

$$\begin{aligned} y_n &= \sum_{n\in\mathbb{Z}} h_k(z^k x_n) \\ &= \left(\sum_{n\in\mathbb{Z}} h_k z^k\right) x_n, \end{aligned}$$

that is, symbolically,

$$y_n = H(z)x_n. \tag{123}$$

In some cases (see the examples below) a function $H(z)$ holomorphic in a *ring* $\{r_1 < |z| < r_2\}$ *containing the unit circle* $\{|z| = 1\}$ is given. This function defines a convolutional filter whose impulse response h_n is given by the Laurent expansion (see [B6], Theorem 1.22, p. 53)

$$H(z) = \sum_{n\in\mathbb{Z}} h_n z^n \qquad (r_1 < |z| < r_2). \tag{124}$$

In particular, the Laurent expansion at $z = 1$ is absolutely convergent, and thus the impulse response h_n is stable. The frequency response of the filter is

$$\tilde{h}(\omega) = H(e^{-i\omega}). \tag{125}$$

Recall that the Laurent expansion is explicitly given by the Cauchy formula

$$h_n = \frac{1}{2i\pi}\oint_C \frac{H(z)}{z^{n+1}}\,\mathrm{d}z, \tag{126}$$

where C is a closed path without multiple points that lies within the interior of the ring of convergence, for example the unit circle, taken in the anti-clockwise sense. The method of residues can be used to compute the right-hand side of formula (126). This equality also takes the form

$$h_n = \frac{1}{2\pi}\int_0^{2\pi} H(e^{-i\omega})e^{in\omega}\,d\omega\,. \tag{127}$$

The integral in (126) can also, have been computed by the method of residues: If C is a simple closed contour on which f is analytic, except for a finite number of isolated singular points $z_1, \dots, z_N$, then

$$\oint_C f(z)\,dz = 2i\pi \sum_{k=1}^{N} a_k,$$

where a_k is the residue of f at $z = z_k$ (see [B1], Chapter 4, pp. 207 and following). This is the *Cauchy residue theorem* . In the case where f has a pole of order m at $z = z_k$, the residue at this point is given by formula

$$a_k = \frac{1}{(m-1)!}\frac{d^{m-1}}{dz^{m-1}}[f(z)(z-z_k)^{m-1}]|_{z=z_k}.$$

EXERCISE **B3.4.** *Compute*

$$\oint_C \frac{3z+1}{z(z-1)^3}\,dz.$$

Series, Parallel, and Feedback Configurations

We now describe the basic operations on digital filters (see Fig. B3.1).

Let $\mathcal{L}_1$ and $\mathcal{L}_2$ be two convolutional filters with (stable) impulse responses h_1^n and h_n^2 and transfer functions $H_1(z)$ and $H_2(z)$, respectively.

The *series* filter $\mathcal{L} = \mathcal{L}_2 * \mathcal{L}_1$ is, by definition, the convolutional filter with impulse response $h_n = (h^1 * h^2)_n$ and transfer function $H(z) = H_1(z)H_2(z)$. It operates as follows: The input x_n is first filtered by $\mathcal{L}_1$, and the output of $\mathcal{L}_1$ is then filtered by $\mathcal{L}_2$, to produce the final output y_n.

The *parallel* filter $\mathcal{L} = \mathcal{L}_1 + \mathcal{L}_2$ is, by definition, the convolutional filter with impulse response $h_1^n + h_n^2$ and frequency response $H(z) = H_1(z) + H_2(z)$. It operates as follows: The input x_n is filtered by $\mathcal{L}_1$, and "in parallel," it is filtered by $\mathcal{L}_2$, and the two outputs are added to produce the final output y_n.

The *feedback* filter $\mathcal{L} = \mathcal{L}_1/(1 - \mathcal{L}_1 * \mathcal{L}_2)$ is, by definition, the convolutional filter with impulse response frequency response

$$H(z) = \frac{H_1(z)}{1 - H_1(z)H_2(z)}.$$

This filter will be a convolutional filter if and only if this frequency response is the FT of a stable impulse response.

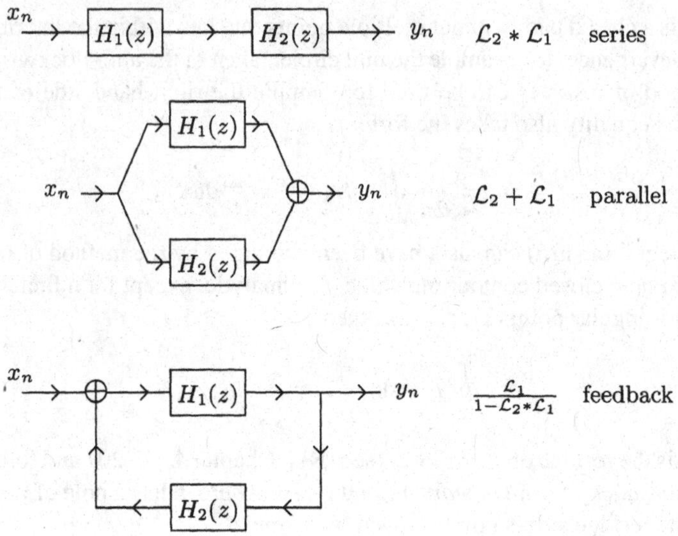

Figure B3.1. Series, parallel, and feedback configurations

The filter $\mathcal{L}_1$ is the *forward loop* filter, whereas $\mathcal{L}_1$ is the *feedback loop* filter. The forward loop processes the total input, which consists of the input x_n plus the fed-back input, that is, the output y_n processed by the feedback loop filter.

EXERCISE **B3.1.** *Consider the filter with impulse response*

$$h_n = \left(\frac{1}{3}\right)^n 1_{\{n\geq 0\}}.$$

Give a feedback representation of this filter.

Rational Transfer Functions

Let

$$P(z) = 1 + \sum_{j=1}^{p} a_j z^j, \qquad Q(z) = 1 + \sum_{\ell=1}^{q} b_\ell z^\ell \tag{128}$$

be two polynomials with complex coefficients of the complex variable z. We shall assume that $P(z)$ has no roots on the unit circle $\{|z| = 1\}$.

Let $r_1 = \max\{|z| : P(z) = 0 \text{ and } |z| < 1\}$; $r_1 = 0$ if there is no root of $P(z)$ with modulus strictly smaller than 1.

Let $r_2 = \inf\{|z| : P(z) = 0 \text{ and } |z| > 1\}$; $r_2 = +\infty$ if there is no root of $P(z)$ with modulus strictly larger than 1.

The function

$$H(z) = \frac{Q(z)}{P(z)} \tag{129}$$

is holomorphic in the ring $C_{r_1,r_2} = \{r_1 < |z| < r_2\}$ (in the open disk $\{|z| < r_2\}$ if $r_1 = 0$) which contains the unit circle since $r_2 > 1$. We thus have a Laurent expansion in C_{r_1,r_2}

$$H(z) = \sum_{n\in\mathbb{Z}} h_n z^n, \tag{130}$$

which defines a filter with *stable* impulse response h_n and frequency response

$$H(e^{i\omega}) = \frac{Q(e^{-i\omega})}{P(e^{-i\omega})}$$

(see [B1], Section 3.3, or [B6], Theorem 1.22, p. 153).

EXAMPLE **B3.1.** *An integer $r \geq 1$ and $\gamma \in \mathbb{C}$ are given. We set*

$$H(z) = \frac{1}{(z-\gamma)^r}.$$

First Case: $|\gamma| > 1$. The ring of convergence is defined by $r_2 = |\gamma|$ and $r_1 = 0$ (thus, in fact we have a disk of convergence $\{|z| < |\gamma|\}$ that contains the unit circle). The Laurent expansion is in this case a power-series expansion in the neighborhood of zero

$$H(z) = \sum_{n\geq 0} h_n z^n \qquad (|z| < |\gamma|).$$

To find the impulse response h_n we must expand $H(z)$ as a power series. But $(-1)^{r-1}(r-1)!(z-\gamma)^{-r}$ is the $(r-1)$st derivative of

$$\frac{1}{z-\gamma} = -\frac{1}{\gamma}\left(1 + \frac{z}{\gamma} + \frac{z^2}{\gamma^2} + \cdots + \frac{z^n}{\gamma^n} + \cdots\right), \quad |z| < \gamma,$$

and therefore, for $|z| < \gamma$,

$$(-1)^{r-1}(r-1)!(z-\gamma)^{-r-1}$$

$$= -\frac{1}{\gamma}\sum_{n=r-1}^{\infty} n(n-1)\ldots(n-r+2)\frac{z^{n-r+1}}{\gamma^n},$$

$$= -\frac{1}{\gamma}\sum_{j=0}^{\infty}\frac{(j+r-1)!}{j!}\frac{z^j}{\gamma^j}\frac{1}{\gamma^{r-1}}.$$

Finally,

$$\frac{1}{(z-\gamma)^r} = \frac{(-1)^r}{\gamma^r(r-1)!}\sum_{j=0}^{\infty}\frac{(j+r-1)!}{j!\gamma^j}z^j,$$

and, identifying this expression with $\sum_{j=0}^{\infty} h_j z^j$, we obtain

$$h_n = (-1)^r\frac{(n+r-1)!}{n!(r-1)!}\left(\frac{1}{\gamma}\right)^{r+n}, \qquad n \geq 0,$$

with $h_n = 0$ if $n < 0$.

Second Case: $|\gamma| < 1$. *The Laurent expansion is then a power-series expansion in the neighborhood of* ∞:

$$\frac{1}{(z-\gamma)^r} = \sum_{n\geq 0} h_{-n} z^{-n}.$$

Changing z *into* $1/\zeta$,

$$\left(\frac{1}{\zeta}-\gamma\right)^{-r} = (-1)^r \frac{\zeta^r}{\gamma^r}\left(\zeta-\frac{1}{\gamma}\right)^{-r}, \quad |\zeta| < \frac{1}{|\gamma|}.$$

We can use the previous calculations to obtain

$$\left(\frac{1}{\zeta}-\gamma\right)^{-r} = H(\zeta^{-1}) = \frac{1}{\gamma^r (r-1)!}\sum_{j=r}^{\infty}(j-1)\ldots(j-r+1)\gamma^j\zeta^j,$$

and we obtain the anticausal filter

$$h_{-n} = \frac{(n-1)!}{(r-1)!(n-r)!}\gamma^{-n-r}, \quad n \geq r,$$

where $h_n = 0$ *if* $n > -r$.

Linear Recurrence Equations

If y_n is the output of the filter with transfer function (129) corresponding to the stable input signal x_n, we have $\tilde{y}(\omega) = H(e^{-i\omega})\tilde{x}(\omega)$, that is,

$$P(e^{-i\omega})\tilde{y}(\omega) = Q(e^{-i\omega})\tilde{x}(\omega).$$

Now $P(e^{-i\omega})\tilde{y}(\omega)$ is the Fourier sum of the signal $y_n + \sum_{j=1}^{p} a_j y_{n-j}$, and $Q(e^{-i\omega})\tilde{x}(\omega)$ is the Fourier sum of $x_n + \sum_{\ell=1}^{q} b_\ell x_{n-\ell}$. Therefore,

$$y_n + \sum_{j=1}^{p} a_j y_{n-j} = x_n + \sum_{\ell=1}^{q} b_\ell x_{n-\ell}, \tag{131}$$

or, symbolically,

$$P(z)y_n = Q(z)x_n.$$

The general solution of the recurrence equation (131) is the sum of an arbitrary solution and of the general solution of the equation without right-hand side

$$y_n + \sum_{j=1}^{p} a_j y_{n-j} = 0.$$

This latter equation has for a general solution a weighted sum of terms of the form

$$r(n)\rho^{-n},$$

where ρ is a root of $P(z)$ and $r(n)$ is a polynomial of degree equal to the multiplicity of this root minus one. If we are given x_n, $n \in \mathbb{Z}$, and the initial conditions $y_0, y_{-1}, \ldots, y_{-p+1}$, the solution of (131) is completely determined.

In order that the general solution never blows up (it is said to blow up if $\lim_{|n|\uparrow\infty} |y_n| = \infty$) whatever the stable input x_n, $n \in \mathbb{Z}$, and for any initial conditions $y_{-p+1}, \ldots, y_{-1}, y_0$, it is necessary and sufficient that all the roots of $P(z)$ have modulus strictly greater than unity.

A particular solution of (131) is

$$y_n = \sum_{k\geq 0} h_k x_{n-k} .$$

The output y_n is stable when the input x_n is stable since the impulse response h_n is itself stable, and therefore y_n does not blow up.

Therefore, we see that in order for the general solution of (131) with stable input x_n to be stable, it is necessary and sufficient that the polynomial $P(z)$ has all its roots with modulus strictly greater than 1.

DEFINITION **B3.1.** *The rational filter $Q(z)/P(z)$ is said to be* stable and causal *if $P(z)$ has all its roots outside the closed unit disk* $\{|z| \leq 1\}$.

Causality arises from the property that if $P(z)$ has roots with modulus strictly greater than unity $Q(z)/P(z) = H(z)$ is analytic inside $\{|z| < r_2\}$ where $r_2 > 1$. The Laurent expansion of $H(z)$ is then an expansion as an entire series $H(z) = \sum_{k\geq 0} h_k z^k$, and this means that the filter is causal ($h_k = 0$ when $k < 0$).

DEFINITION **B3.2.** *The stable rational filter $Q(z)/P(z)$ is said to be* causally invertible *if $Q(z)$ has all its roots outside the closed unit disk* $\{|z| \leq 1\}$.

In fact, writing the analytic expansion of $P(z)/Q(z)$ in the neighborhood of zero as $\sum_{k\geq 0} w_k z^k$, we have

$$\tilde{x}(\omega) = \frac{P(e^{-i\omega})}{Q(e^{-i\omega})}\,\tilde{y}(\omega) = \left(\sum_{k\geq 0} w_k e^{-ik\omega}\right)\tilde{y}(\omega),$$

that is,

$$x_n = \sum_{k\geq 0} w_k y_{n-k} . \tag{132}$$

B3·3 All-Pass and Spectral Factorization

All-Pass Filters

A particular case of a rational filter is the all-pass filter.

THEOREM **B3.3.** *Let z_i $(1 \leq i \leq L)$ be complex numbers with modulus strictly greater than* 1. *Then the transfer function*

$$H(z) = \prod_{i=1}^{L} \frac{zz_i^* - 1}{z - z_i} \tag{133}$$

satisfies

$$|H(z)| \begin{cases} < 1 & \text{if } |z| < 1, \\ = 1 & \text{if } |z| = 1, \\ > 1 & \text{if } |z| > 1. \end{cases} \tag{134}$$

Proof: Let

$$H_i(z) = \frac{zz_i^* - 1}{z - z_i}$$

be an arbitrary factor of $H(z)$. If $|z| = 1$, we observe that $|H_i(z)| = 1$, using *Féjer's identity*

$$(z - \beta)(z - \frac{1}{\beta^*}) = -\frac{1}{\beta^*} z|z - \beta|^2, \tag{135}$$

which is true for $|z| = 1$, $\beta \in \mathbb{C}$, $\beta \neq 0$. On the other hand, $H_i(z)$ is holomorphic on $|z| < |z_i|$ and $|H_i(0)| = |z_i|^{-1} < 1$. Therefore, we must have $|H_i(z)| < 1$ on $\{|z| < |z_i|\}$, otherwise the maximum modulus theorem for holomorphic functions would be contradicted. (Recall the maximum modulus theorem: If f is analytic in a bounded region D and $|f|$ is continuous in the closure of D, then $|f|$ takes its maximum on the boundary of D; see [B1], Theorem 2.66, p. 97, or [B6], Theorem 1.21, p. 51.) Observing that

$$\left| H_i\left(\frac{1}{z^*}\right) \right| = \frac{1}{|H_i(z)|}$$

we see that the result just obtained implies that $|H_i(z)| > 1$ if $|z| > 1$. ■

A filter with frequency response $H(e^{-i\omega})$ is a pure phase filter, or *all-pass filter*, by definition. It is called all-pass because its gain is unity: $|H(e^{-i\omega})| = 1$.

Consider a signal x_n such that

$$x_n = \begin{cases} a_n, & 0 \leq n \leq N, \\ 0 & \text{otherwise.} \end{cases}$$

It can be represented by its polynomial z-transform

$$A(z) = \sum_{n=0}^{N} a_n z^n.$$

Let $z_1, z_2, \ldots, z_N$ be its roots. In particular,

$$A(z) = a_N \prod_{j=1}^{N} (z - z_j).$$

The effect of filtering x_n with an all-pass filter $(z_1^* z - 1)/(z - z_1)$ is to replace the factor $z - z_1$ in $A(z)$ by $z_1^* z - 1$, but it does not change the energy of the signal.

Indeed, the z-transform of the resulting signal,

$$B(z) = A(z)\,\frac{z_1^* z - 1}{z - z_1},$$

is such that

$$|B(e^{-i\omega})|^2 = |A(e^{-i\omega})|^2,$$

and therefore,

$$\sum_{n=0}^{N} |b_n|^2 = \frac{1}{2\pi}\int_{-\pi}^{+\pi} |B(e^{-i\omega})|^2\, d\omega$$

$$= \frac{1}{2\pi}\int_{-\pi}^{+\pi} |A(e^{-i\omega})|^2\, d\omega = \sum_{n=0}^{N} |a_n|^2.$$

Thus,

$$\sum_{n=0}^{N} |a_n|^2 = \sum_{n=0}^{N} |b_n|^2. \tag{136}$$

At a time $0 \le k \le N$ the two signals $(a_0, \dots, a_N)$ and $(b_0, \dots, b_N)$ have already dissipated the energies

$$E_a(k) = \sum_{j=0}^{k} |a_j|^2 \quad \text{and} \quad E_b(k) = \sum_{j=0}^{k} |b_j|^2.$$

There is an interesting relation between these partial energies. Writing

$$A(z) = (z - z_1)F(z), \qquad B(z) = (z_1^* z - 1)F(z),$$

where

$$F(z) = f_0 + f_1 z + \cdots + f_{N-1} z^{N-1},$$

we have

$$a_n = f_{n-1} - z_1 f_n, \quad b_n = z_1^* f_{n-1} - f_n \quad (0 \le n \le N),$$

where $f_{-1} = f_N = 0$ by convention. Taking the square of the modulus and subtracting yields

$$|a_n|^2 - |b_n|^2 = (|z_1|^2 - 1)(|f_n|^2 - |f_{n-1}|^2),$$

and therefore,

$$E_a(k) - E_b(k) = (|z_1|^2 - 1)|f_k|^2. \tag{137}$$

This shows that if $|z_1| < 1$, then $(a_0, \dots, a_n)$ is always late with respect to $(b_0, \dots, b_N)$ in dissipating its energy.

Féjer's Lemma

EXERCISE **B3.2.** *Let x_n be a stable signal with z-transform $X(z)$. Define its autocorrelation function c_n by*

$$c_n = \sum_{k\in\mathbb{Z}} x_{n+k}x_k^*.$$

Show that $\{c_n\}_{n\in\mathbb{Z}} \in \ell^2_{\mathbb{C}}(\mathbb{Z})$ and that its Fourier sum is

$$\tilde{c}(\omega) = |\tilde{x}(\omega)|^2 = R(e^{-i\omega}),$$

where

$$R(z) = X(z)X(z)^*.$$

The 2π-periodic function $R(e^{-i\omega})$ in the above exercise has the following properties:

$$R(e^{-i\omega}) \geq 0, \tag{138}$$

$$\int_{-\pi}^{+\pi} R(e^{-i\omega}) < \infty. \tag{139}$$

Moreover, if $X(z)$ is a rational fraction,

$$R(e^{-i\omega}) \text{ is a rational fraction in } e^{-i\omega}. \tag{140}$$

The next result is *Féjer's lemma*, which is also called the *spectral factorization* theorem.

THEOREM **B3.1.** *Let $R(z)$ be a rational fraction in z with complex coefficients such that* (138) *and* (139) *are satisfied. Then there exist two polynomials in z with complex coefficients, $P(z)$ and $Q(z)$, and a constant $c \geq 0$, such that $P(0) = Q(0) = 1$ and*

$$R(e^{-i\omega}) = c\left|\frac{Q(e^{-i\omega})}{P(e^{-i\omega})}\right|^2. \tag{141}$$

Moreover, one can choose $P(z)$ to be without roots inside the closed unit disk, and $Q(z)$ to be without roots inside the open unit disk.

Proof: $R(z)$ can be factored as

$$R(z) = az^{m_0}\prod_{k\in K}(z - z_k)^{m_k},$$

where $a \in \mathbb{C}$, the z_k are nonnull distinct complex numbers, and the $m_k \in \mathbb{Z}$. If $|z| = 1$, $R(z)$ is real, and therefore,

$$R(z) = R(z)^* = a^*(z^*)^{m_0}\prod_{k\in K}(z^* - z_k^*)^{m_k} = a^*(z^{-1})^{m_0}\prod_{k\in K}(z^{-1} - z_k^*)^{m_k}.$$

Therefore, when $|z| = 1$, there exist $b \in \mathbb{C}$ and $r_0 \in \mathbb{Z}$ such that

$$R(z) = bz^{r_0}\prod_{k\in K}\left(z - \frac{1}{z_k^*}\right)^{m_k}.$$

Therefore, if $|z| = 1$,

$$az^{m_0} \prod_{k \in K} (z - z_k)^{m_k} = bz^{r_0} \prod_{k \in K} \left(z - \frac{1}{z_k^*} \right)^{m_k}.$$

Two rational fractions that coincide when $|z| = 1$ coincide for all $z \in \mathbb{C}$. In particular, $a = b$, and whenever we have in $R(z)$ the factor $(z - z_k)$ with $|z_k| \neq 1$, then we also have the factor $(z - \frac{1}{z_k^*})$. We therefore have

$$\prod_{k \in K} (z - z_k)^{m_k} = \prod_{j \in J} (z - z_j)^{s_j} \left(z - \frac{1}{z_j^*} \right)^{s_j} \prod_{\ell \in L} (z - z_\ell)^{r_\ell},$$

where $|z_\ell| = 1$ for all $\ell \in L$, and $|z_j| \neq 1$ for all $j \in J$. We show that $r_\ell = 2s_\ell \in \mathbb{N}$ for all $\ell \in L$. For this, we write $z_\ell = e^{-i\omega_\ell}$, and observe that in the neighborhood of ω_ℓ, $R(e^{-i\omega})$ is equivalent to a constant times $(\omega - \omega_\ell)^{r_\ell}$ and therefore can remain nonnegative if and only if $r_\ell = 2s_\ell$. Since $R(e^{-i\omega})$ is locally integrable, then necessarily $s_\ell \in \mathbb{N}$. Therefore,

$$R(z) = bz^{r_0} \prod_{j \in J} (z - z_j)^{s_j} \left(z - \frac{1}{z_j^*} \right)^{s_j} \prod_{\ell \in L} (z - z_\ell)^{2s_\ell}.$$

Using Féjer's identity (135), we therefore find that $R(z)$ can be put under the form

$$R(z) = cz^d |G(z)|^2,$$

where

$$G(z) = \prod_{j \in J} (z - z_j)^{s_j} \prod_{\ell \in L} (z - z_\ell)^{s_\ell}.$$

The function $R(e^{-i\omega})$ can remain real and nonnegative if and only if $c \geq 0$ and $d = 0$. Finally, we can always suppose that $|z_j| < 1$ for all $j \in J$ (a root z_j is paired with another root $1/z_j^*$). ■

EXERCISE **B3.3.** *Find a constant c and polynomials $P(z)$ and $Q(z)$ as in Theorem B3.1, such that*

$$\frac{5 - 2\cos(\omega)}{3 - \cos(\omega)}) = c \left| \frac{Q(e^{-i\omega})}{P(e^{-i\omega})} \right|^2.$$

The proof of Theorem B3.1 can be specialized to obtain that for any polynomial $p(z)$ such that $p(e^{-i\omega}) \geq 0$ for all $\omega \in \mathbb{R}$, there exists a polynomial $A(z)$ with $A(0) = 1$ and no root inside the closed unit disk, and a constant $c \geq 0$, such that

$$p(e^{-i\omega}) = c|A(e^{-i\omega})|^2.$$

Looking at the proof of B3.1, we see if there exist another polynomial $B(z)$ with $B(0) = 1$ and a constant $c' \geq 0$, such that $p(e^{-i\omega}) = c'|B(e^{-i\omega})|^2$, then $c = c'$ and

$$B(z) = H(z)A(z)$$

for some all-pass filter $H(z)$.

B4

Subband Coding

B4·1 Band Splitting with Perfect Reconstruction

Smooth Filter Banks

Let $x(t)$ be a stable base-band (B) real signal that we seek to *analyze* in the following sense. For fixed $N = 2^k$ we wish to obtain for all $1 \leq i \leq 2^k$ the signals $x_i(t)$ with Fourier transforms

$$\hat{x}_i(\nu) = 1_{B_i}(\nu)\hat{x}(\nu),$$

where B_i is the frequency band

$$B_i = \left[\frac{i-1}{2^k}B, \frac{i}{2^k}B\right].$$

From a theoretical point of view the problem is stated with its solution: For each i, do no more than filter $x(t)$ with a pass-band filter of frequency response $1_{B_i}(\nu)$! From the practical point of view of digital processing, in the sample domain, an ideal band-pass filter has an *infinite impulse response*—actually one with rather slow decay—and this makes the above pure band-pass filters of poor value from a numerical point of view.

A solution consists of replacing the pure band-pass filters by approximations with "good" impulse responses, and if possible *finite impulse responses* (FIR). However, FIR filters with short impulse response have in general a poor frequency resolution, and therefore the analysis will not be satisfactory without a careful choice of the approximate band-pass filters. One also requires perfect synthesis, that is,

$$x(t) = \sum_{i=1}^{2^k} \tilde{x}_i(t),$$

where $\tilde{x}_i(t)$ is obtained from $x(t)$ by approximate band-pass filtering on the band B_i. This means that leakage between contiguous bands must be mutually compensated.

The above is a summary of the numerical problem associated with subband decomposition of a signal by a filter bank. The second problem is algorithmic: How to perform efficiently analysis and synthesis? The standard example of an efficient algorithm is the FFT, which involves successive splitting, and subband decomposition is another avatar of this idea: The basic block of the algorithm consists of splitting a given band in two, that is, of solving the subband decomposition problem for $N = 2$.

Subband coding is one way of performing *data compression*. Instead of sampling the original signal and then quantifying the resulting samples with a view of digitizing them, one performs the sampling and quantifying operations on each of the outputs $x_i(t)$. If a subband B_i is deemed unimportant it will be allocated fewer compression resources, that is, only coarsely quantified. The appraisal of the importance of each subband is generally based on psychological experiments. The subjective difference between subbands is very marked in two-dimensional signal processing, where it has been observed that low-frequency components are the most important from a subjective point of view.

The Basic Algorithm

Since all signals and filters considered in the present chapter are real, we need only consider positive frequencies, those in the frequency band $[0, B]$. Ideal splitting of the frequency band $[0, B]$ uses two ideal band-pass filters, one for the band $[0, B/2]$ and the other for the band $[B/2, B]$. We call $T_0(\nu)$ and $T_1(\nu)$ their frequency responses. Then, as the Shannon–Nyquist theorem suggests, we sample each output at rate B, and reconstruction is performed by two ideal band-pass filters, $[0, B/2]$ and $[B/2, B]$, respectively. We call $\widetilde{T}_0(\nu)$ and $\widetilde{T}_1(\nu)$ their frequency responses (of course, if we use ideal pass-band filters, $\widetilde{T}_0(\nu) = T_0(\nu)$, and $\widetilde{T}_1(\nu) = T_1(\nu)$; we keep different notations because in the nonideal case, the analysis and reconstruction filters need not be the same).

Consider Fig. B4.1. In the ideal case (ideal pass-band filters), the signals in the upper branch at levels α ($x_1(t)$) and γ ($y_1(t)$) are identical and equal to the original signal $x(t)$ filtered by the band-pass $[0, B/2]$. This follows from the theory of sampling of Chapter B2, and the details of the operations in the lower branch are shown in Figure B4.2. Similarly, in the lower branch of Fig. B4.1, the signals at levels α ($x_2(t)$) and γ ($y_2(t)$) are identical and equal to the original signal $x(t)$ filtered by the band-pass $[B/2, B]$.

As we explained before, the ideal band-pass filters will be replaced by approximations $T_0(\nu)$ and $\widetilde{T}_0(\nu)$ that have most of their energy inside the band $[0, B/2]$,

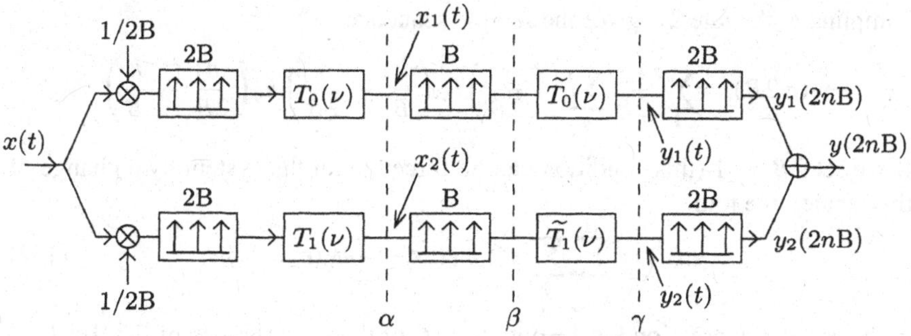

Figure B4.1. Block diagram of subband coding

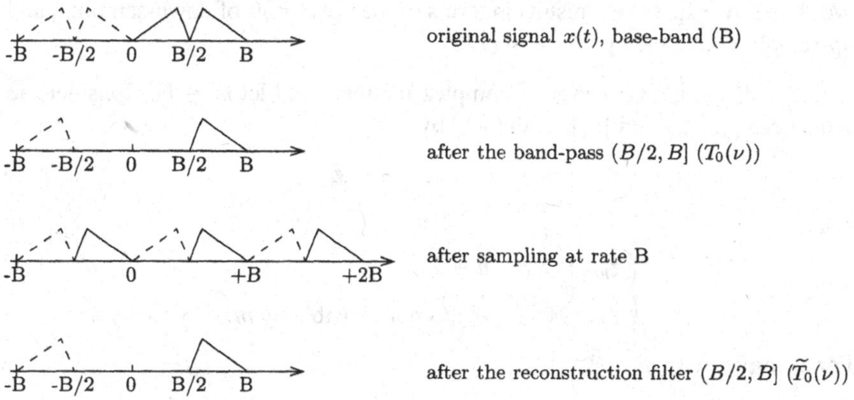

Figure B4.2. Subband coding in the frequency domain

and $T_1(\nu)$ and $\widetilde{T}_1(\nu)$ that have most of theirs inside $[B/2, B]$. We insist once more on the fact that we do not require that $\widetilde{T}_0(\nu) = T_0(\nu)$ nor that $\widetilde{T}_1(\nu) = T_1(\nu)$, because we need some freedom in the choice of $\widetilde{T}_0(\nu)$ and $\widetilde{T}_1(\nu)$ to guarantee perfect reconstruction. *Analysis* of the original signal yields the decomposition $(x_1(t), x_2(t))$, whereas *synthesis* reconstructs $y(t) = x_1(t) + x_2(t)$. Synthesis is called perfect when $y(t) = x(t)$.

The signal at level α is

$$x_1(t) = \frac{1}{2B} \sum_{j \in \mathbb{Z}} x\left(\frac{j}{2B}\right) h_0\left(t - \frac{j}{2B}\right),$$

where $h_0(t)$, $h_1(t)$, $\tilde{h}_0(t)$, $\tilde{h}_1(t)$ are the respective impulse responses corresponding to the frequency responses $T_0(\nu)$, $\widetilde{T}_0(\nu)$, $T_1(\nu)$, $\widetilde{T}_1(\nu)$. Similarly the signal $y_1(t)$ at level γ in Fig. B4.1 is

$$y_1(t) = \sum_k x_1\left(\frac{k}{B}\right) \tilde{h}_0\left(t - \frac{k}{B}\right).$$

Sampling at the rate $2B$ gives the sample sequence

$$y_1(n\,2B) = \sum_k \frac{1}{2B} \sum_j x\left(\frac{j}{2B}\right) h_0\left(\frac{k}{B} - \frac{j}{2B}\right) \tilde{h}_0\left(\frac{n}{2B} - \frac{k}{B}\right).$$

If we set $2B = 1$ (this condition can be forced upon the system by a change of time scale), we find

$$y_1(n) = \sum_k \sum_j x(j) h_0(2k - j) \tilde{h}_0(n - 2k), \tag{142}$$

with a similar expression for the output $y_2(t)$ of the lower branch of Fig. B4.1.

Down- and Up-sampling

We shall now express the results in terms of the operations of down-sampling and up-sampling, and then go back to (142).

Let $\{x_n\}_{n\in\mathbb{Z}}$ be a sequence of complex numbers and let $m \in \mathbb{N}$. Consider the sequences $\{y_n\}_{n\in\mathbb{Z}}$ and $\{z_n\}_{n\in\mathbb{Z}}$ defined by

$$y_n = x_{nm}, \quad n \in \mathbb{Z}$$

and

$$\begin{cases} z_{nm} = x_n, & n \in \mathbb{Z}, \\ z_j = 0 & \text{if } j \text{ is not divisible by } m. \end{cases}$$

For example, with $m = 2$,

$$\begin{array}{cccccc} x_0 & x_1 & x_2 & x_3 & x_4 & x_5 \\ \| & & \| & & \| & \\ y_0 & & y_1 & & y_2 & \end{array}$$

and

$$\begin{array}{ccccccccccc} x_0 & 0 & x_1 & 0 & x_2 & 0 & x_3 & 0 & x_4 & 0 & x_5 \\ \| & \| & \| & \| & \| & \| & \| & \| & \| & \| & \| \\ z_0 & z_1 & z_2 & z_3 & z_4 & z_5 & z_6 & z_7 & z_8 & z_9 & z_{10} \end{array}$$

The sequence $\{y_n\}_{n\in\mathbb{Z}}$ is said to be obtained from the original sequence $\{x_n\}_{n\in\mathbb{Z}}$ by down-sampling by a factor m. The corresponding operation is denoted as $m\downarrow$. Up-sampling by a factor m, denoted as $m\uparrow$, is the operation that transforms $\{x_n\}_{n\in\mathbb{Z}}$ into $\{z_n\}_{n\in\mathbb{Z}}$.

In this chapter, we are concerned with the case $m = 2$. For future use, we shall express the operation of down-sampling by 2 followed by up-sampling by 2 in terms of z-transforms (see Figure B4.3).

Denote $X(z)$ and $R(z)$ the z-transforms of the sequences $\{x(n)\}_{n\in\mathbb{Z}}$ and $\{r(n)\}_{n\in\mathbb{Z}}$, respectively. The sequence $\{r(n)\}_{n\in\mathbb{Z}}$ is therefore obtained from $\{x(n)\}_{n\in\mathbb{Z}}$ by

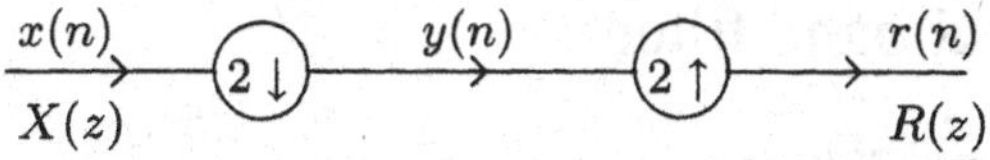

Figure B4.3. Down-sampling and up-sampling

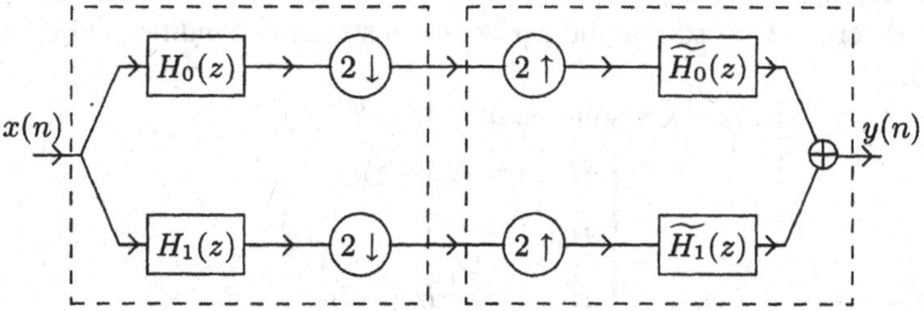

Figure B4.4. Subband coding in the Z-domain (1 split)

replacing all the entries with an odd index by a zero. Therefore,

$$R(z) = \sum_{n\in\mathbb{Z}} x(2n)z^{2n} = \frac{1}{2}\left\{\sum_{n\in\mathbb{Z}} x(n)z^n + \sum_{n\in\mathbb{Z}} x(n)(-z)^n\right\},$$

that is,

$$R(z) = \tfrac{1}{2}\{X(z) + X(-z)\}. \tag{143}$$

Going back to (142) and the similar expression for the lower branch of Fig. B4.1, we see that the whole system is equivalent in the z-domain to Fig. B4.4.

From (143) we see that

$$Y(z) = \tfrac{1}{2}\{X(z)H_0(z) + X(-z)H_0(-z)\}\widetilde{H}_0(z)$$
$$+ \tfrac{1}{2}\{X(z)H_1(z) + X(z)H_1(-z)\}\widetilde{H}_1(z).$$

Separating the aliasing terms from the rest,

$$Y(z) = \tfrac{1}{2}X(z)\{H_0(z)\widetilde{H}_0(z) + H_1(z)\widetilde{H}_1(z)\}$$
$$+ \tfrac{1}{2}X(-z)\{H_0(-z)\widetilde{H}_0(z) + H_1(-z)\widetilde{H}_1(z)\}. \tag{144}$$

Therefore, aliasing is eliminated if

$$H_0(-z)\widetilde{H}_0(z) + H_1(-z)\widetilde{H}_1(z) = 0, \tag{145}$$

and perfect reconstruction is obtained provided that

$$H_0(z)\widetilde{H}_0(z) + H_1(z)\widetilde{H}_1(z) = 2. \tag{146}$$

B4·2 FIR subband filters

Quadrature Mirrors Filters

In order to find a solution of (145) and (146), one can first fix H_0 and then find H_1, $\widetilde{H}_0$, $\widetilde{H}_1$ in terms of H_0 in order to satisfy the no-aliasing condition (145). Then one can determine H_0 so that the perfect reconstruction conditions (146) can be satisfied.

Given $H_0(z)$, one possible solution of (145) is[6]

$$\begin{cases} H_1(z) = H_0(-z), \\ \widetilde{H}_0(z) = H_0(z), \\ \widetilde{H}_1(z) = -H_0(-z). \end{cases} \tag{147}$$

Assume that the filter H_0 is symmetric, that is, it has a symmetric impulse response ($h_0(-n) = h_0(n)$, $n \in \mathbb{Z}$). Then

$$\begin{aligned} H_1(z) = H_0(-z) &= \sum_{n\in\mathbb{Z}} (-1)^n h_0(n) z^n \\ &= \sum_{n\in\mathbb{Z}} (-1)^n h_0(-n) z^n \quad \text{(symmetry of } H_0) \\ &= \sum_{n\in\mathbb{Z}} (-1)^n h_0(n) \left(\frac{1}{z}\right)^n . \end{aligned}$$

Therefore, if H_0 is symmetric,

$$H_1(z) = H_0\left(-\frac{1}{z}\right),$$

that is, in terms of pulsations,

$$H_1(e^{-i\omega}) = H_0(e^{-i(\pi-\omega)}).$$

This means that the pulsation spectrum of H_1 is symmetric with respect to that of H_0 with respect to the frequency $\pi/2$. This is why in this case H_0 and H_1 are said to be *quadrature mirror filters* (QMFs).

Going back to (147)—and without assuming that H_0 is symmetric—the perfect reconstruction condition (146) becomes, in terms of H_0:

$$H_0(z)^2 - H_0(-z)^2 = 2. \tag{148}$$

One drawback of the solution (145) is the nonexistence of a finite impulse response filter H_0 satisfying it. However, we can relax condition (146) to

$$H_0(z)\widetilde{H}_0(z) + H_1(z)\widetilde{H}_1(z) = 2z^k \tag{149}$$

---

[6]Esteban, D., and Galand, C. (1977), Applications of quadrature mirror filters to split-band voice-coding schemes, *Proc. IEEE Int. Conf. ASSP*, Hartford, Connecticut, 191–195.

for some $K \geq 1$, which means that we accept a delay of K time units to recover the input, and in this case FIR filters do exist.

EXAMPLE **B4.1.** *Taking the no-aliasing condition* (147) *into account, the relaxed condition* (149) *with* $K = 1$ *gives*

$$H_0(z)^2 - H_0(-z)^2 = 2z. \tag{150}$$

A famous solution is the Haar filter

$$H_0(z) = \frac{1}{\sqrt{2}}(1+z). \tag{151}$$

The relaxed condition (149) allows a "linear phase" corresponding to a delay K. For $K \geq 2$, we shall just mention that (149) does not have an exact solution with a FIR filter.

If in Fig. B4.1, the input signal $x(t)$ is assumed to be band-pass $[jB, (j+1)B]$, for some $j \geq 1$, the resulting output of the analyzer is the same as if the input had been frequency-shifted by jB, to obtain a base-band (B) signal. In fact, immediately after the sampler at rate B, at level γ, we have the same signal, for both inputs (the pass-band signal $x(t)$, or its base-band version).

Therefore, the analyzer of Fig. B4.1 performs band splitting on the base-band (B) version of any band-pass $[jB, (j+1)B]$ signal. Consequently, the analyzer of Fig. B4.4 behaves in the same way, with the additional feature of being *independent* of B!

This remark shows that the full program of subband coding can be achieved by a cascade of the analysis (resp., synthesis) structures of Fig. B4.4 (by anticipation, let us mention that this is similar to the structure of Mallat's algorithm in multiresolution analysis). Fig. B4.5 shows the analysis synthesis of the band $[0, B]$ into four subbands $[0, B/4]$, $[B/4, B/2]$, $[B/2, 3B/4]$, $[3B/4, B]$.

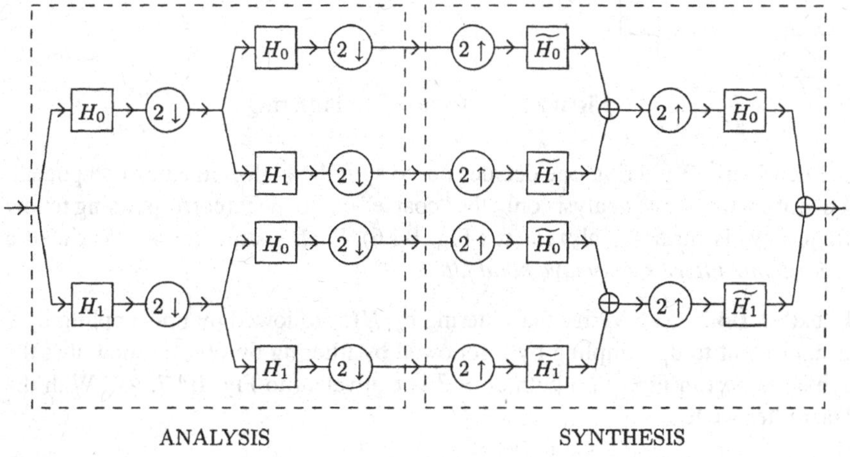

Figure B4.5. Subband coding in the Z-domain (2 splits)

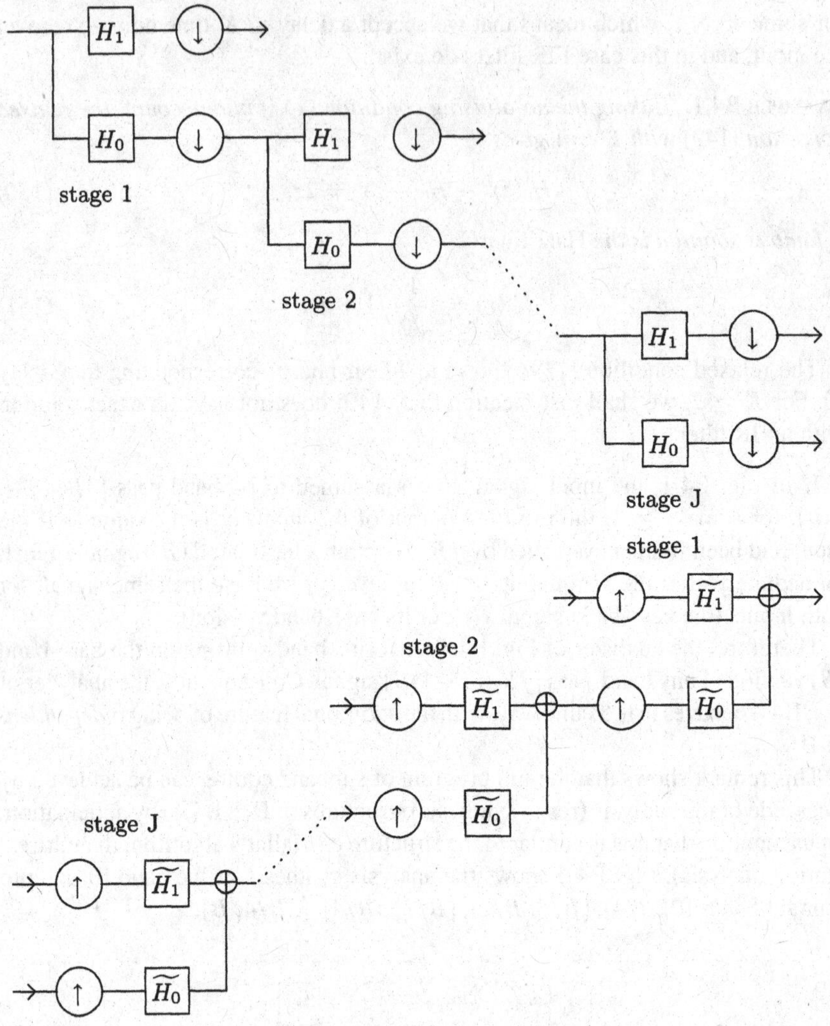

Figure B4.6. Octave band filtering

In constant Q-filtering one decides not to split the high-frequency component. Thus at each stage of analysis only the "coarse" component (corresponding to low frequency) is further analyzed (see Fig. B4.6). Such a structure is also called a *logarithmic filter* or an *octave band filter*.

EXERCISE **B4.2.** (a): Verify that filtering by $H(z)$ followed by up-sampling by 2 is equivalent to up-sampling by 2 followed by filtering by $H(z^2)$. Show that the synthesis part in Fig. B4.6 with $J = 3$ is equivalent to Fig. B4.7. (b): With the Haar filter we have

$$\widetilde{H}_0(z) = \frac{1}{\sqrt{2}}(1+z), \quad \widetilde{H}_1(z) = \frac{1}{\sqrt{2}}(1-z).$$

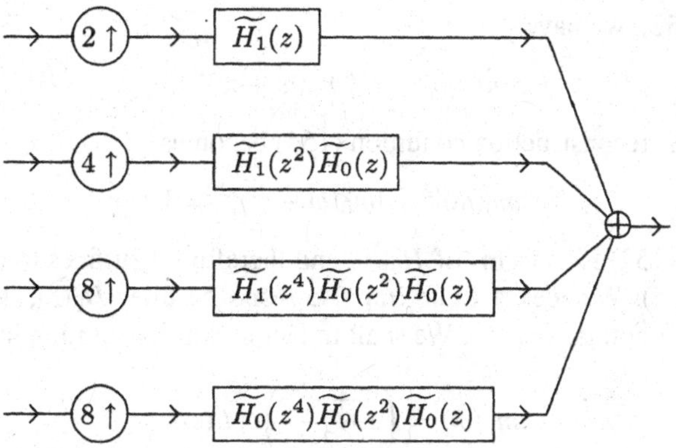

Figure B4.7. Equivalent octave band filtering

Give the impulse response of each of the four filters in Fig. B4.7.

Another Solution

Another class of solutions[7] for the no-aliasing condition (145) is

$$\begin{cases} H_1(z) = z^{-1} H_0(-z^{-1}), \\ \widetilde{H}_0(z) = H_0(z^{-1}), \\ \widetilde{H}_1(z) = z H_0(-z). \end{cases} \tag{152}$$

The perfect reconstruction condition (146) then becomes

$$H_0(z)H_0(z^{-1}) + H_0(-z)H_0(-z^{-1}) = 2. \tag{153}$$

Since H_0 is a real filter,

$$H_0(z^{-1}) = H(z)^* \quad \text{for } z = e^{-i\omega},$$

and (153) takes the form

$$|H_0(e^{-i\omega})|^2 + |H_0(-e^{-i\omega})|^2 = 2. \tag{154}$$

We shall now exhibit a general solution of (152). We perform a change of notation that will be convenient in the chapter on multiresolution analysis:

$$H_0(z) = H(z) = \sum_{n \in \mathbb{Z}} h_n z^n,$$

$$m_0(\omega) = \frac{1}{\sqrt{2}} H_0(e^{-i\omega}),$$

$$m_1(\omega) = \frac{1}{\sqrt{2}} H_1(e^{-i\omega}).$$

---

[7]Smith, M.J.T., and Barnwell, III T.P. (1986), Exact reconstruction techniques for tree-structured subband coders, *IEEE Transactions ASSP*, **34**, 434–441.

In view of (152), we have

$$m_1(\omega) = e^{i\omega} m_0(\omega + \pi)^*,$$

and the perfect reconstruction condition (154) becomes

$$|m_0(\omega)|^2 + |m_0(\omega + \pi)|^2 = 1. \tag{155}$$

The solution (152) is in terms of $H_0(z)$, and therefore it suffices to obtain $m_0(\omega)$ satisfying (155). We seek a finite impulse response filter $H_0(z)$, in which case $m_0(\omega)$ is a polynomial in $e^{-i\omega}$. We shall in fact look for a solution in the form

$$m_0(\omega) = \left(\frac{1 + e^{i\omega}}{2}\right)^N L(\omega),$$

where $N \geq 1$, and $L(\omega)$ is a polynomial in $e^{-i\omega}$. Letting

$$M_0(\omega) = |m_0(\omega)|^2,$$

we have

$$M_0(\omega) = \left|\frac{1 + e^{i\omega}}{2}\right|^{2N} |L(\omega)|^2 = \left(\cos^2\left(\frac{\omega}{2}\right)\right)^N |L(\omega)|^2.$$

But $|L(\omega)|^2$ is a *real-valued* polynomial in $e^{-i\omega}$, and therefore it is a polynomial in $\cos(\omega)$. Since $\cos(\omega) = 1 - 2\sin^2(\omega/2)$,

$$M_0(\omega) = \left(\cos^2\left(\frac{\omega}{2}\right)\right)^N P\left(\sin^2\left(\frac{\omega}{2}\right)\right),$$

for some polynomial P. Condition (155) must be satisfied for all ω, and therefore it is equivalent to

$$(1 - y)^N P(y) + y^N P(1 - y) = 1 \tag{156}$$

for all $y \in [0, 1]$. Since two polynomials identical on $[0, 1]$ are identical everywhere, the latter equality is for all $y \in \mathbb{R}$.

The polynomials y^N and $(1 - y)^N$ have no common roots, and therefore, by Bezout's theorem, there exist two unique polynomials a and b, of degree $\leq N - 1$, such that

$$(1 - y)^N a(y) + y^N b(y) = 1. \tag{157}$$

This is true for all $y \in \mathbb{R}$, and in particular, replacing y by $1 - y$,

$$(1 - y)^N b(1 - y) + y^N a(1 - y) = 1.$$

By the uniqueness of a and b, it follows that

$$b(y) = a(1 - y).$$

Therefore, (157) is

$$(1 - y)^N a(y) + y^N a(1 - y) = 1.$$

Therefore, $P(y) = a(y)$ is a solution of (156). We have thuse proven that (156) admits at least one solution, and by the uniqueness in Bezout's theorem, this solution is the only one of degree $\leq N-1$. We have

$$a(y) = (1-y)^{-N}[1 - y^N a(1-y)] = \sum_{k=0}^{N-1} \binom{N+k-1}{k} y^k + O(y^N).$$

Since a is a polynomial of degree $\leq N-1$, it is equal to its Taylor series truncated at $N-1$, and therefore,

$$a(y) = \sum_{k=0}^{N-1} \binom{N+k-1}{k} y^k.$$

This solution is the unique one with degree $\leq N-1$. Observe that it is nonnegative for all $y \in [0, 1]$, and therefore a solution to the initial problem. Call it P_N and let P be the general solution. We have

$$(1-y)^N(P(y) - P_N(y)) + y^N(P(1-y) - P_N(1-y)) = 0.$$

This implies that $P - P_N$ is divisible by y^N, that is, $P(y) - P_N(y) = y^N Q(y)$, and

$$(1-y)^N y^N Q(y) + y^N(1-y)^N Q(1-y)) = 0,$$

which implies $Q(y) + Q(1-y)) = 0$. That is, Q is symmetric with respect to $1/2$, and therefore of the form $Q(y) = R(1/2 - y)$ for an odd polynomial R.

In summary, the general solution of (156) is

$$P(y) = \sum_{k=0}^{N-1} \binom{N+k-1}{k} y^k + y^N R\left(\frac{1}{2}-\right), \tag{158}$$

where $R(y)$ is any odd polynomial such that $P(y)$ so defined remains nonnegative for all $y \in [0, 1]$.

Having obtained $M_0(\omega)$, it remains to extract its square root $m_0(\omega)$. But this can be done by spectral factorization, using Féjer's lemma.

We shall close this chapter on the basic principles of subband coding. Note, however, that other solutions were proposed, most notably "biorthogonal solutions,"[8] which are more versatile and yield finite impulse response subband filters with better properties (of symmetry, for instance). We refer to the monograph [B12], where the reader will find a full and detailed treatment of this topic, as well as additional references.

[8]Vetterli, M. *Filter banks allowing perfect reconstruction*, Signal Processing, 10 (3), 1986, 219–244.

References

[B1] Ablowitz, M.J. and Jokas, A.S. (1997). *Complex Variables*, Cambridge University Press.

[B2] Daubechies, I. (1992). *Ten Lectures on Wavelets, CBSM–NSF Regional Conf. Series in Applied Mathematics*, SIAM: Philadelphia, PA.

[B3] Gasquet, C. and Witomski, P. (1991). *Analyse de Fourier et Applications*, Masson: Paris.

[B4] Haykin, S. (1989). *An Introduction to Analog and Digital Communications*, Wiley: New York.

[B5] Hirsch, M.W. and Smale, S. (1974). *Differential Equations, Dynamical Systems, and Linear Algebra*, Academic Press: San Diego.

[B6] Kodaira, K. (1984). *Introduction to Complex Analysis*, Cambridge University Press.

[B7] Lighthill, M.J. (1980). *An Introduction to Fourier Analysis and Generalized Functions*, Cambridge University Press.

[B8] Nussbaumer, H.J. (1981). *Fast Fourier Transform and Convolution Algorithms*, Springer–Verlag: New York.

[B9] Orfanidis, S. (1985). *Optimal Signal Processing*, McMillan: New York.

[B10] Papoulis, A. (1984). *Signal Analysis*, McGraw-Hill: New York.

[B11] Rudin, W. (1966) *Real and Complex Analysis*, McGraw-Hill: New York.

[B12] Vetterli, M. and Kovačević, J. (1995). *Wavelets and Sub-Band Coding*, Prentice-Hall: Englewood Cliffs, NJ.

Part C

Fourier Analysis in L^2

Introduction

The modern era of Fourier theory started when the tools of functional analysis—in particular, Lebesgue's integral and Hilbert spaces—became available. Fourier theory then seemed to have reached the promised land, which is called L^2, the space of square-integrable complex functions, indeed a Hilbert space.

F. Riesz and E. Fischer were the first to study Fourier series in the L^2 framework.[1] Many ideas of the modern theory of Hilbert spaces were already contained in the work of these two mathematicians, and they had a clear view of the geometric aspect of the L^2-spaces. They were inspired by a series of articles by David Hilbert written after 1904 on the theme of integral equations and in which he gives the properties of $\ell^2_{\mathbb{C}}(\mathbb{Z})$. Note, however, that the notion of abstract Hilbert spaces made its appearance much later than one usually believes, in the years 1927–1930, with the work of John von Neumann, who was motivated by quantum mechanics.[2]

In short, a Hilbert space is a vector space H on the field $\mathbb{C}$ (or $\mathbb{R}$), with a Hermitian (or scalar) product, denoted $\langle \cdot, \cdot \rangle$ or $\langle \cdot, \cdot \rangle_H$, and a special topological property that we shall now briefly introduce. The Hermitian product induces a norm, the norm $\|x\|$, or $\|x\|_H$, of the vector $x \in H$ being

$$\|x\| = \langle x, x \rangle^{\frac{1}{2}}.$$

[1] F. Riesz, Sur les systèmes orthonormaux de fonctions, *CRAS Paris*, 144, 1907, 615–619; and E. Fischer, Sur la convergence en moyenne, *CRAS Paris*, 144, 1907, 1022–1024; Applications d'un théorème sur la convergence en moyenne, *CRAS Paris*, 144, 1907, 1148–1151.

[2] His theory was published in the reference text *Mathematische Grundlagen der Quantum Mechanik* in 1932.

This allows us to define a limit in H: We say that $\lim_{n\to\infty} x_n = x$ if $\lim_{n\to\infty} \|x_n - x\| = 0$. Having this notion of a limit, we have the notion of a Cauchy sequence: A sequence $\{x_n\}_{n\in\mathbb{N}}$ in H is called a Cauchy sequence if

$$\lim_{m,n\to\infty} \|x_m - x_n\| = 0.$$

To be called a Hilbert space, H must—besides being a vector space on $\mathbb{C}$ with a Hermitian product—be complete with respect to the induced norm. This means that any Cauchy sequence $\{x_n\}_{n\in\mathbb{N}}$ in H converges, that is, there exists an $x \in H$ such that

$$\lim_{n\to\infty} \|x_n - x\| = 0.$$

Note that for any positive integer k, $\mathbb{C}^k$, considered as a vector space on $\mathbb{C}$ with the usual Hermitian product, is indeed a Hilbert space. But there are more sophisticated Hilbert spaces. For instance, $L^2_{\mathbb{C}}(\mathbb{R})$, the space of functions $f : \mathbb{R} \to \mathbb{C}$ that are square-integrable:

$$\int_{\mathbb{R}} |f(t)|^2 \, dt < \infty.$$

In $L^2_{\mathbb{C}}(\mathbb{R})$, one does not distinguish two functions that are almost everywhere equal. The Hermitian product is

$$\langle f, g \rangle = \langle f, g \rangle_{L^2_{\mathbb{C}}(\mathbb{R})} = \int_{\mathbb{R}} f(t) g(t)^* \, dt.$$

Saying that $\lim_{n\uparrow\infty} f_n = f$ in $L^2_{\mathbb{C}}(\mathbb{R})$ means that

$$\lim_{n\uparrow\infty} \int_{\mathbb{R}} |f_n(t) - f(t)|^2 \, dt = 0.$$

Another example of a Hilbert space is the space of functions $f : \mathbb{R} \to \mathbb{C}$ that are 2π-periodic, and in $L^2_{\mathbb{C}}([-\pi, +\pi])$, that is, square-integrable on $[-\pi, +\pi]$:

$$\int_{-\pi}^{+\pi} |f(t)|^2 \, dt < \infty,$$

with the Hermitian product

$$\langle f, g \rangle = \langle f, g \rangle_{L^2_{\mathbb{C}}([-\pi,+\pi])} = \int_{-\pi}^{+\pi} f(t) g(t)^* \, dt.$$

In $L^2_{\mathbb{C}}([-\pi, +\pi])$, one also does not distinguish two functions that are almost everywhere equal.

A third example is $\ell^2_{\mathbb{C}}(\mathbb{Z})$, the set of complex sequences $a = \{x_n\}_{n\in\mathbb{Z}}$ such that

$$\sum_{n\in\mathbb{Z}} |x_n|^2 < \infty,$$

with the Hermitian product

$$\langle a, b\rangle = \langle a, b\rangle_{\ell^2_{\mathbb{C}}(\mathbb{Z})} = \sum_{n\in\mathbb{Z}} a_n b_n^*.$$

The Hilbert space $L^2_{\mathbb{C}}(\mathbb{R})$ is a paradise of Fourier transforms, since every function thereof admits a Fourier transform, and moreover the mapping that associates to a function its Fourier transform is a bijection from $L^2_{\mathbb{C}}(\mathbb{R})$ to itself, and the inversion formula for Fourier transforms, which gives the latter in terms of the former is

$$f(t) = \int_{\mathbb{R}} \hat{f}(\nu)\, e^{+2i\pi\nu t}\, \mathrm{d}\nu.$$

This is not a precise statement. In particular, the integrals appearing in the definition of the transform and in the inversion formula are in some extended sense, and the equality in the inversion formula is "almost everywhere." To be exact,

$$f(t) = \lim_{a,b\uparrow\infty} \int_{-a}^{b} \hat{f}(\nu)\, e^{+2i\pi\nu t}\, \mathrm{d}\nu,$$

where the limit is in the sense of $L^2_{\mathbb{C}}(\mathbb{R})$. A similar interpretation is needed for the integral defining the Fourier transform.

The beautiful formula of the L^2-theory is the Plancherel–Parseval's formula

$$\int_{\mathbb{R}} \hat{f}(\nu)\hat{g}(\nu)^*\, \mathrm{d}\nu = \int_{\mathbb{R}} f(t)g(t)^*\, \mathrm{d}t,$$

in other words,

$$\langle f, g\rangle_{L^2_{\mathbb{C}}(\mathbb{R})} = \langle \hat{f}, \hat{g}\rangle_{L^2_{\mathbb{C}}(\mathbb{R})},$$

where $f, g \in L^2_{\mathbb{C}}(\mathbb{R})$.

The above results are stated for the Fourier integral transform, but similar results hold for the Fourier series of periodic functions: Let f be a 2π-periodic function square-integrable on $[0, 2\pi]$; then it admits the representation

$$f(t) = \sum_{n\in\mathbb{Z}} c_n(f)e^{int}.$$

This is the inversion formula for Fourier series. Similarly to the Fourier transform in $L^2_{\mathbb{C}}(\mathbb{R})$, this equality is only almost everywhere, and the sum has to be interpreted in an extended sense:

$$f(t) = \lim_{N\uparrow\infty} \sum_{-N}^{+N} c_n(f)e^{int},$$

where the limit is in the sense of $L^2_{\mathbb{C}}([-\pi, +\pi])$. This result is in fact a particular case of the Hilbert basis theorem, which gives the orthonormal expansion

$$x = \sum_{n\in\mathbb{Z}} \langle x, e_n\rangle e_n$$

of any vector x in a Hilbert space H, when $\{e_n\}_{n\in\mathbb{N}}$ is a complete orthonormal system. "Orthonormal" means that

$$\langle e_k, e_n\rangle = 1_{k=n},$$

and "complete" means that the closure of the vector space consisting of the finite linear combination of elements of $\{e_n\}_{n\in\mathbb{N}}$ is H. (See Chapter C2 for precise definitions.) In this case, the above orthonormal expansion is valid (the series in the right-hand side converging with respect to the distance induced by the Hermitian product of H), and moreover, we have Plancherel–Parseval's identity

$$\|x\|^2 = \sum_{n\in\mathbb{Z}} \|\langle x, e_n\rangle\|^2.$$

The Fourier series development is a particular case of the above very general result, where $H \equiv L^2_{\mathbb{C}}([-\pi, +\pi])$, and

$$e_n(t) \equiv \frac{1}{\sqrt{2\pi}} e^{int}.$$

The Plancherel–Parseval formula for Fourier series reads

$$\frac{1}{2\pi}\int_{-\pi}^{+\pi} f(t)g(t)^*\,dt = \sum_{n\in\mathbb{Z}} c_n(f)c_n(g)^*,$$

where f and g are 2π-periodic functions in $L^2_{\mathbb{C}}([-\pi, +\pi]$. In terms of Hermitian products,

$$\frac{1}{2\pi}\langle f, g\rangle_{L^2_{\mathbb{C}}([-\pi,+\pi])} = \langle c(f), c(g)\rangle_{\ell^2_{\mathbb{C}}(\mathbb{Z})}.$$

C1
Hilbert Spaces

C1·1 Basic Definitions

Hilbert space theory is the fundamental tool in Fourier analysis of finite-energy signals. It is a huge chapter of functional analysis, but we shall only give the definitions and prove the basic facts used in this book, in particular, the projection theorem and the theorem of extension of isometries.

Pre-Hilbert Spaces

DEFINITION **C1.1.** *Let E be a vector space over $\mathbb{C}$ (resp., $\mathbb{R}$) and let $(x, y) \to \langle x, y\rangle$ be a mapping from $E \times E$ to $\mathbb{C}$ (resp., $\mathbb{R}$) such that, for all $x, y \in E$ and all $\alpha \in \mathbb{C}$ (resp., $\mathbb{R}$),*

(a) $\langle x + z, y\rangle = \langle x, y\rangle + \langle z, y\rangle$,
(b) $\langle \alpha x, y\rangle = \alpha\langle x, y\rangle$,
(c) $\langle x, y\rangle = \langle y, x\rangle^*$ *(resp.,* $\langle x, y\rangle = \langle y, x\rangle$*),*
(d) $\langle x, x\rangle \geq 0$, *and* $\langle x, x\rangle = 0$ *if and only if* $x = 0$.

It is then said that E is a complex (resp., real) pre-Hilbert space *with the* Hermitian product *(resp., scalar product)* $\langle \cdot, \cdot\rangle$. *The complex (resp., real) number* $\langle x, y\rangle$ *is the Hermitian (resp., scalar) product of x and y.*

In the above definition and in the sequel, 0 represents the zero of $\mathbb{R}$ or $\mathbb{C}$, or the neutral element of addition in E. The context will remove ambiguity.

From now on, we shall consider *complex* pre-Hilbert (and later Hilbert) spaces. The other choice for the scalar field, $\mathbb{R}$, leads to formally analogous results.

For any $x \in E$, denote

$$\|x\|^2 = \langle x, x\rangle. \tag{1}$$

Elementary computations yield

$$\|x+y\|^2 = \|x\|^2 + \|y\|^2 + 2\,\mathrm{Re}\,\{\langle x, y\rangle\} \tag{2}$$

for any $x, y \in E$. The *parallelogram identity*

$$\|x\|^2 + \|y\|^2 = \tfrac{1}{2}(\|x+y\|^2 + \|x-y\|^2) \tag{3}$$

is obtained by expanding the right-hand side of (3) using (2).

EXERCISE **C1.1.** *Prove the* polarization identity

$$\langle x, y\rangle = \frac{1}{4}\{\|x+y\|^2 - \|x-y\|^2 + i\|x+iy\|^2 - i\|x-iy\|^2\}. \tag{4}$$

Consequently, two Hermitian products $\langle\cdot,\cdot\rangle_1$ and $\langle\cdot,\cdot\rangle_2$ on E such that $\|\cdot\|_1 = \|\cdot\|_2$ are identical.

THEOREM **C1.1.** *For all $x, y \in E$, we have the* Schwarz inequality

$$|\langle x, y\rangle| \le \|x\| \times \|y\|, \tag{5}$$

with equality holding if and only if there exist $\alpha, \beta \in \mathbb{C}$ such that $\alpha x + \beta y = 0$.

Proof: We do the proof for the real case and leave the complex case to the reader. We may assume that $\langle x, y\rangle \neq 0$; otherwise, the result is trivial. For all $\lambda \in \mathbb{R}$,

$$\|x\|^2 + 2\lambda\langle x, y\rangle^2 + \lambda^2\langle x, y\rangle^2\|y\|^2 = \|x + \lambda\langle x, y\rangle y\|^2 \ge 0.$$

This second-degree polynomial in $\lambda \in \mathbb{R}$ therefore cannot have two distinct real roots, and this implies a nonpositive discriminant:

$$|\langle x, y\rangle|^4 - \|x\|^2\,|\langle x, y\rangle|^2\|y\|^2 \le 0,$$

and thus the inequality (5) holds. Equality in (5) corresponds to a null discriminant, and this in turn implies a double root λ of the polynomial. For such a root, $\|x + \lambda\langle x, y\rangle y\|^2 = 0$, that is, by Property (d) in Definition C1.1,

$$x + \lambda\langle x, y\rangle y = 0. \qquad \blacksquare$$

DEFINITION **C1.2.** *Two elements $x, y \in E$ are said to be* orthogonal *if $\langle x, y\rangle = 0$.*

Let $x_1, \ldots, x_n \in E$ be pairwise orthogonal. We have Pythagoras' theorem:

$$\left\|\sum_{i=1}^{n} x_i\right\|^2 = \sum_{i=1}^{n}\|x_i\|^2, \tag{6}$$

which follows from (2).

THEOREM **C1.2.** *The mapping $x \to \|x\|$ is a norm on E, that is, for all $x, y \in E$, all $\alpha \in \mathbb{C}$,*

(a) $\|x\| \ge 0$, and $\|x\| = 0$ if and only if $x = 0$,

(b) $\|\alpha x\| = |\alpha|\, \|x\|$,
(c) $\|x + y\| \le \|x\| + \|y\|$.

Proof: The proof of (a) and (b) is immediate. For (c) write

$$\|x + y\|^2 = \|x\|^2 + \|y\|^2 + \langle x, y\rangle + \langle y, x\rangle$$

and

$$(\|x\| + \|y\|)^2 = \|x\|^2 + \|y\|^2 + 2\|x\|\,\|y\|.$$

It therefore suffices to prove

$$\langle x, y\rangle + \langle y, x\rangle \le 2\|x\|\,\|y\|,$$

and this follows from the Schwarz inequality. ■

The norm $\|\cdot\|$ induces a distance $d(\cdot,\cdot)$ on E by

$$d(x, y) = \|x - y\|. \tag{7}$$

Recall that a mapping $d : E \times E \to \mathbb{R}_+$ is called a distance on E if, for all $x, y, z \in E$,

(a′) $d(x, y) \ge 0$, and $d(x, y) = 0$ if and only if $x = y$,
(b′) $d(x, y) = d(y, x)$,
(c′) $d(x, y) \ge d(x, z) + d(z, y)$.

The above properties are immediate consequences of (a), (b), and (c) of Theorem C1.2.

Hilbert Spaces

A *metric space* is a set E endowed with a distance d. One then says: the metric space (E, d), or, for short and when the context is sufficiently explicit as to the choice of the distance, the metric space E. A pre-Hilbert space E is therefore a metric space for the distance d induced by the Hermitian product.

DEFINITION **C1.3.** *A* Hilbert space *is a pre-Hilbert space that is complete with respect to the distance d defined above.*

Recall that a metric space (E, d) is called *complete* if any *Cauchy sequence* in E converges; that is, if $\{x_n\}_{n\ge1}$ is a sequence in E such that $\lim_{m,n\uparrow\infty} d(x_n, x_m) = 0$, then there exists $x \in E$ such that $\lim_{n\uparrow\infty} d(x_n, x) = 0$.

When considered as a Hilbert space relative to the norm $\|\cdot\|$, E will be denoted H. If necessary, the notation for the Hermitian product and the norm will explicitly refer to the space H: We then write $\langle\cdot,\cdot\rangle_H$ and $\|\cdot\|_H$.

EXAMPLE **C1.1.** *Let $(X, \mathcal{X}, \mu)$ be a measure space. It is proven in the appendix (Theorem 26) that $L^2_{\mathbb{C}}(\mu)$ is a Hilbert space relative to the Hermitian product*

$$\langle f, g\rangle_{L^2_{\mathbb{C}}(\mu)} = \int_X f(x)g(x)^*\,\mu(dx).$$

EXAMPLE **C1.2.** *A particular case of Example 3.1 is that in which $X = \mathbb{Z}$, $\mathcal{X} = \mathcal{P}(\mathbb{Z})$, and μ is the counting measure on $\mathbb{Z}$. The corresponding Hilbert space $L^2_{\mathbb{C}}(\mu)$ is then denoted $\ell^2_{\mathbb{C}}(\mathbb{Z})$. Therefore,*

$$\ell^2_{\mathbb{C}}(\mathbb{Z}) = \{\{x_n\}_{n\in\mathbb{Z}} : x_n \in \mathbb{C} \text{ for all } n \in \mathbb{Z} \text{ and } \sum_{n\in\mathbb{Z}} |x_n|^2 < \infty\}$$

is a Hilbert space with Hermitian product

$$\langle x, y\rangle_{\ell^2_{\mathbb{C}}(\mathbb{Z})} = \sum_{n\in\mathbb{Z}} x_n y_n^*.$$

EXERCISE **C1.2.** *Show that if $h(t)$ and $x(t)$ are both in $L^2_{\mathbb{C}}(\mathbb{R})$, then $y = h * x$ is well defined. Find $h \in L^2_{\mathbb{C}}(\mathbb{R})$ such that $\|h\| = 1$ and maximizing $y(T)$ for a given time T. What is the corresponding maximum?*

C1·2 Continuity Properties

Closed Subspaces

A subset G is said to be closed in H if every convergent sequence of G has a limit in G.

THEOREM **C1.3.** *Let $G \subset H$ be a vector subspace of the Hilbert space H. Endow G with the Hermitian product that is the restriction to G of the Hermitian product on H. Then G is a Hilbert space if and only if G is closed in H.*

In this case, G is called a Hilbert subspace *of H.*

Proof: Assume that G is closed. Let $\{x_n\}_{n\in\mathbb{N}}$ be a Cauchy sequence in G. It is a fortiori a Cauchy sequence in H, and therefore it converges in H to some x, and this x must be in G, because it is a limit of elements of G and G is closed.

Assume that G is a Hilbert space with the Hermitian product induced by the Hermitian product of H. In particular, every convergent sequence $\{x_n\}_{n\in\mathbb{N}}$ of elements of G converges to some element of G. Therefore, G is closed. ■

EXERCISE **C1.3.** *Let $(X, \mathcal{X}, \mu)$ be a measure space. For some fixed constant K, let $G = L^2_{\mathbb{C}}(\mu) \cap \{f; \sup|f(t)| \le K,\ \mu - a.e.\}$. Is G a Hilbert subspace of $L^2_{\mathbb{C}}(\mu)$? Answer the same question, with $G = L^2_{\mathbb{C}}(\mu) \cap \{f; \sup|f(t)| < \infty,\ \mu - a.e.\}$. What about $L^2_{\mathbb{C}}(\mu) \cap \{f; \sup|f(t)| \le K(f) < \infty,\ \mu - a.e.\}$?*

DEFINITION **C1.4.** *Let $\{x_t\}_{t\in T}$ be an arbitrary collection of elements of H. The smallest Hilbert subspace of H containing all the vectors x_t, $t \in T$, is called the* Hilbert subspace generated by *$\{x_t\}_{t\in T}$, or the* Hilbert span *of $\{x_t\}_{t\in T}$.*

EXERCISE **C1.4.** *Call G be the Hilbert subspace generated by $\{x_t\}_{t\in T}$. Let $\mathcal{L}$, the vector space formed by all finite linear combinations of elements of $\{x_t\}_{t\in T}$. Show that $G = \overline{\mathcal{L}}$ (the closure of $\mathcal{L}$).*

The following notation is convenient:

$$\mathcal{L} = \text{span}\,\{x_t, t \in T\},$$
$$G = \overline{\text{span}}\,\{x_t, t \in T\}.$$

Paraphrasing the above result, we see that $x \in \overline{\text{span}}\,\{x_t, t \in T\}$ if and only if x is the limit in H of a sequence of finite linear combinations of elements of $\{x_t, t \in T\}$.

Continuity of the Hermitian product

THEOREM **C1.4.** *Let H be a Hilbert space over $\mathbb{C}$ with the Hermitian product $\langle \cdot, \cdot \rangle$. The mapping from $H \times H$ into $\mathbb{C}$ defined by $(x, y) \mapsto \langle x, y \rangle$ is bicontinuous.*

Proof: We have

$$|\langle x + h_1, y + h_2 \rangle - \langle x, y \rangle| = |\langle x, h_2 \rangle + \langle h_1, y \rangle + \langle h_1, h_2 \rangle|.$$

By Schwarz's inequality, $|\langle x, h_2 \rangle| \le \|x\|\|h_2\|$, $|\langle h_1, y \rangle| \le \|y\|\|h_1\|$, and $|\langle h_1, h_2 \rangle| \le \|h_1\|\|h_2\|$. Therefore,

$$\lim_{\|h_1\|, \|h_2\| \downarrow 0} |\langle x + h_1, y + h_2 \rangle - \langle x, y \rangle| = 0. \qquad \blacksquare$$

In particular, the norm $x \mapsto \|x\|$ is a continuous function from H to $\mathbb{R}_+$.

EXERCISE **C1.5.** *Let $(X, \mathcal{X}, \mu)$ be a measure space, where μ is a* finite *measure. Let $\{f_n\}n \ge 1$ be a sequence of $L^2_{\mathbb{C}}(\mu)$ converging to f. Apply Theorem C1.4 to prove that $\lim_{n\uparrow\infty} \mu(f_n) = \mu(f)$. Give a counterexample of this property when the hypothesis that μ is finite is dropped. (Hint: $f = 1_{[0,1]}$, $f_n = (1-1/n)1_{[0,1]} + \cdots$.) Show that when μ is finite,*

$$G = L^2_{\mathbb{C}}(\mu) \cap \{f; \mu(f) = 0\}$$

is a Hilbert subspace of $L^2_{\mathbb{C}}(\mu)$.

Note that when μ is not finite, G need not be a Hilbert subspace of $L^2_{\mathbb{C}}(\mu)$. Wavelet multiresolution analysis will provide a spectacular counterexample.

Isometry Extension Theorem

DEFINITION **C1.5.** *Let H and K be two Hilbert spaces with Hermitian products denoted $\langle \cdot, \cdot \rangle_H$ and $\langle \cdot, \cdot \rangle_K$, respectively, and let $\varphi : H \mapsto K$ be a linear mapping such that, for all $x, y \in H$,*

$$\langle \varphi(x), \varphi(y) \rangle_K = \langle x, y \rangle_H \,. \tag{8}$$

Then φ is called a linear isometry *from H into K. If, moreover, φ is from H* onto *K, then H and K are said to be* isomorphic.

Note that a linear isometry is necessarily injective, since $\varphi(x) = \varphi(y)$ implies $\varphi(x - y) = 0$, and therefore,

$$0 = \|\varphi(x - y)\|_K = \|x - y\|_H \,,$$

and this implies $x = y$. In particular, if the linear isometry is *onto*, it is necessarily bijective.

Recall that a subset $A \in E$, where (E, d) is a metric space, is said to be *dense* in E if, for all $x \in E$, there exists a sequence $\{x_n\}_{n\geq 1}$ in A converging to x. The following result will often be used. It is called the isometry extension theorem of Hilbert spaces or, for short, the *isometry extension theorem.*

THEOREM **C1.5.** *Let H and K be two Hilbert spaces with Hermitian products $\langle \cdot, \cdot \rangle_H$ and $\langle \cdot, \cdot \rangle_K$, respectively. Let V be a vector subspace of H that is dense in H, and let $\varphi : V \mapsto K$ be a linear isometry from V to K (φ is linear and* (8) *holds for all $x, y \in V$). Then there exists a unique linear isometry $\widetilde{\varphi} : H \mapsto K$ such that the restriction of $\widetilde{\varphi}$ to V is φ.*

Proof: We shall first define $\widetilde{\varphi}(x)$ for $x \in H$. Since V is dense in H, there exists a sequence $\{x_n\}_{n\geq 1}$ in V converging to x. Since φ is isometric,

$$\|\varphi(x_n) - \varphi(x_m)\|_K = \|x_n - x_m\|_H \quad \text{for all } m, n \geq 1.$$

In particular, $\{\varphi(x_n)\}_{n\geq 1}$ is a Cauchy sequence in K, and it therefore converges to some element of K, which we denote $\widetilde{\varphi}(x)$.

The definition of $\widetilde{\varphi}(x)$ is independent of the sequence $\{x_n\}_{n\geq 1}$ converging to x. Indeed, for another such sequence $\{y_n\}_{n\geq 1}$,

$$\lim_{n\uparrow\infty} \|\varphi(x_n) - \varphi(y_n)\|_K = \lim_{n\uparrow\infty} \|x_n - y_n\|_H = 0.$$

The mapping $\widetilde{\varphi} : H \mapsto K$ so constructed is clearly an extension of φ (for $x \in V$, one can take as the approximating sequence of x the sequence $\{x_n\}_{n\geq 1}$ such that $x_n \equiv x$).

The mapping $\widetilde{\varphi}$ is linear. Indeed, let $x, y \in H$, $\alpha, \beta \in \mathbb{C}$, and let $\{x_n\}_{n\geq 1}$ and $\{y_n\}_{n\geq 1}$ be two sequences in V converging to x and y, respectively. Then $\{\alpha x_n + \beta y_n\}_{n\geq 1}$ converges to $\alpha x + \beta y$. Therefore,

$$\lim_{n\uparrow\infty} \varphi(\alpha x_n + \beta y_n) = \widetilde{\varphi}(\alpha x + \beta y).$$

However,

$$\varphi(\alpha x_n + \beta y_n) = \alpha\varphi(x_n) + \beta\varphi(y_n) \to \alpha\widetilde{\varphi}(x) + \beta\widetilde{\varphi}(y).$$

Therefore, $\widetilde{\varphi}(\alpha x + \beta y) = \alpha\widetilde{\varphi}(x) + \beta\widetilde{\varphi}(y)$.

The mapping $\widetilde{\varphi}$ is isometric since, in view of the bicontinuity of the Hermitian product and of the isometricity of φ,

$$\begin{aligned} \langle \widetilde{\varphi}(x), \widetilde{\varphi}(y) \rangle_K &= \lim_{n\uparrow\infty} \langle \varphi(x_n), \varphi(y_n) \rangle_K \\ &= \lim_{n\uparrow\infty} \langle x_n, y_n \rangle_H = \langle x, y \rangle_H, \end{aligned}$$

where $\{x_n\}_{n\geq 1}$ and $\{y_n\}_{n\geq 1}$ are two sequences in V converging to x and y, respectively. ■

C1·3 Projection Theorem

Let G be a Hilbert subspace of the Hilbert space H. The *orthogonal complement* of G in H, denoted $G^\perp$, is defined by

$$G^\perp = \{z \in H : \langle z, x\rangle = 0 \text{ for all } x \in G\}. \tag{9}$$

Clearly, $G^\perp$ is a vector space over $\mathbb{C}$. Moreover, it is closed in H since if $\{z_n\}_{n\geq 1}$ is a sequence of elements of $G^\perp$ converging to $z \in H$, then, by continuity of the Hermitian product,

$$0 = \lim_{n\uparrow\infty} \langle z_n, x\rangle = \langle z, x\rangle \quad \text{for all } x \in H.$$

Therefore, $G^\perp$ is a Hilbert subspace of H.

Observe that a decomposition $x = y+z$ where $y \in G$ and $z \in G^\perp$ is necessarily unique. Indeed, let $x = y' + z'$ be another such decomposition. Then, letting $a = y - y'$, $b = z - z'$, we have that $0 = a + b$ where $a \in G$ and $b \in G^\perp$. Therefore, in particular, $0 = \langle a, a\rangle + \langle a, b\rangle = \langle a, a\rangle$, which implies that $a = 0$. Similarly, $b = 0$.

Theorem C1.6 (Projection theorem). *Let $x \in H$. There exists a* unique *element $y \in G$ such that $x - y \in G^\perp$. Moreover,*

$$\|y - x\| = \inf_{u\in G} \|u - x\|. \tag{10}$$

Proof: Let $d(x, G) = \inf_{z\in G} d(x, z)$ and let $\{y_n\}_{n\geq 1}$ be a sequence in G such that

$$\begin{aligned} d(x, G)^2 &\leq d(x, y_n)^2 \\ &\leq d(x, G)^2 + \frac{1}{n}. \end{aligned} \tag{*}$$

The parallelogram identity gives, for all $m, n \geq 1$,

$$\|y_n - y_m\|^2 = 2(\|x - y_n\|^2 + \|x - y_m\|^2) - 4\|x - \tfrac{1}{2}(y_m + y_n)\|^2.$$

Since $\frac{1}{2}(y_n + y_m) \in G$,

$$\|x - \tfrac{1}{2}(y_m + y_n)\|^2 \geq d(x, G)^2;$$

therefore,

$$\|y_n - y_m\|^2 \leq 2\left(\frac{1}{n} + \frac{1}{m}\right).$$

The sequence $\{y_n\}_{n\geq 1}$ is thus a Cauchy sequence in G, and it consequently converges to some $y \in G$ since G is closed. Passing to the limit in (*) gives (10).

Uniqueness of y satisfying (10): Let $y' \in G$ be another such element. Then

$$\|x - y'\| = \|x - y\| = d(x, G),$$

and from the parallelogram identity

$$\begin{aligned}\|y - y'\|^2 &= 2\|y - x\|^2 + 2\|y' - x\|^2 - 4\|x - \tfrac{1}{2}(y + y')\|^2 \\ &= 4d(x, G)^2 - 4\|x - \tfrac{1}{2}(y + y')\|^2.\end{aligned}$$

Since $\frac{1}{2}(y + y') \in G$,

$$\|x - \tfrac{1}{2}(y + y')\|^2 \geq d(x, G)^2.$$

Therefore,

$$\|y - y'\|^2 \leq 0,$$

which implies that $\|y - y'\|^2 = 0$ and therefore, $y = y'$.

It now remains to show that $x - y$ is orthogonal to G, that is,

$$\langle x - y, z\rangle = 0 \quad \text{for all } z \in G.$$

This is trivially true if $z = 0$, and we shall therefore assume $z \neq 0$. Because $y + \lambda z \in G$ for all $\lambda \in \mathbb{R}$,

$$\|x - (y + \lambda z)\|^2 \geq d(x, G)^2,$$

that is,

$$\|x - y\|^2 + 2\lambda \operatorname{Re}\{\langle x - y, z\rangle\} + \lambda^2\|z\|^2 \geq d(x, G)^2.$$

Since

$$\|x - y\|^2 = d(x, G)^2,$$

we have

$$-2\lambda \operatorname{Re}\{\langle x - y, z\rangle\} + \lambda^2\|z\|^2 \geq 0 \quad \text{for all } \lambda \in \mathbb{R},$$

which implies $\operatorname{Re}\{\langle x - y, z\rangle\} = 0$. The same type of calculation with $\lambda \in i\mathbb{R}$ (pure imaginary) leads to

$$\operatorname{Im}\{\langle x - y, z\rangle\} = 0.$$

Therefore,

$$\langle x - y, z\rangle = 0.$$

That y is the unique element of G such that $y - x \in G^\perp$ follows from the observation made just before the statement of Theorem C1.6. ■

The element y in Theorem C1.6 is called the *orthogonal projection* of x on G (see Fig. C1.1) and is denoted $P_G(x)$.

Projection Principle

The projection theorem states, in particular, that for any $x \in G$ there is a unique decomposition

$$x = y + z, \qquad y \in G,\ z \in G^\perp, \tag{11}$$

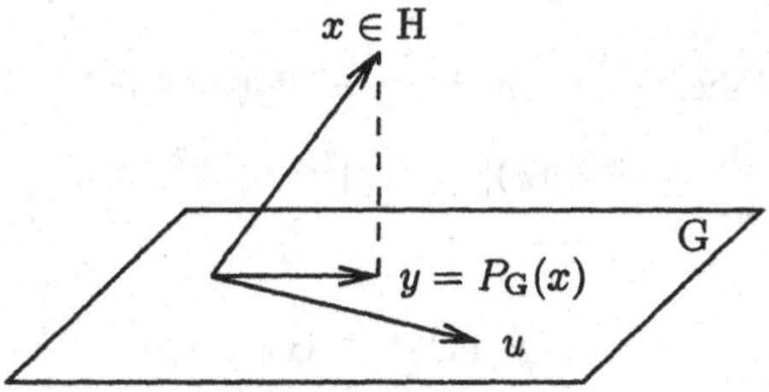

Figure C1.1. Orthogonal projection

and that $y = P_G(x)$, the (unique) element of G closest to x. Therefore, the orthogonal projection $y = P_G(x)$ is characterized by the following two properties:

(1) $y \in G$;
(2) $\langle y - x, z \rangle = 0$ for all $z \in G$.

This characterization is called the *projection principle* and is useful in determining projections.

Projection Operator

The next result features two useful properties of the orthogonal projection operator P_G.

THEOREM **C1.7.** *Let G be a Hilbert subspace of the Hilbert space H.*

(α) The mapping $x \to P_G(x)$ is linear and continuous; furthermore,

$$\|P_G(x)\| \leq \|x\| \quad \textit{for all } x \in H.$$

(β) If F is a Hilbert subspace of H such that $F \subseteq G$, then $P_F \circ P_G = P_F$. In particular, $P_G^2 = P_G$ (P_G is then called idempotent*).*

Proof: (α) Let $x_1, x_2 \in H$. They admit the decomposition

$$x_i = P_G(x_i) + w_i \quad (i = 1, 2),$$

where $w_i \in G^\perp$ $(i = 1, 2)$. Therefore,

$$\begin{aligned} x_1 + x_2 &= P_G(x_1) + P_G(x_2) + w_1 + w_2 \\ &= P_G(x_1) + P_G(x_2) + w, \end{aligned}$$

where $w \in G^\perp$. Now, $x_1 + x_2$ admits a unique decomposition of the type

$$x_1 + x_2 = y + w,$$

where $w \in G^\perp$, $y \in G$: namely, $y = P_G(x_1 + x_2)$. Therefore,

$$P_G(x_1 + x_2) = P_G(x_1) + P_G(x_2).$$

One similarly proves that

$$P_G(\alpha x) = \alpha P_G(x) \quad \text{for all } \alpha \in G,\ x \in H.$$

Thus P_G is linear.

From Pythagoras' theorem applied to $x = P_G(x) + w$,

$$\|P_G(x)\| + \|w\|^2 = \|x\|^2,$$

and therefore,

$$\|P_G(x)\|^2 \le \|x\|^2.$$

Hence, P_G is continuous.

(β) The unique decompositions of x on G and $G^\perp$ and of $P_G(x)$ on F and $F^\perp$ are

$$x = P_G(x) + w,$$

$$P_G(x) = P_F(P_G(x)) + z.$$

From these two equalities we obtain

$$x = P_F(P_G(x)) + z + w. \tag{*}$$

But $(z \in G^\perp) \Rightarrow (z \in F^\perp)$ since $F \subseteq G$, and therefore $v = z + w \in F^\perp$. On the other hand, $P_F(P_G(x)) \in F$. Therefore, (*) is the unique decomposition of x on F and $F^\perp$; in particular, $P_F(x) = P_F(P_G(x))$. ■

The next result says that the projection operator P_G is "continuous" with respect to G.

THEOREM **C1.8.** *(i) Let $\{G_n\}_{n\ge1}$ be a nondecreasing sequence of Hilbert subspaces of H. Then the closure G of $\bigcup_{n\ge1} G_n$ is a Hilbert subspace of H and, for all $x \in H$,*

$$\lim_{n\uparrow\infty} P_{G_n}(x) = P_G(x).$$

(ii) Let $\{G_n\}$ be a nonincreasing sequence of Hilbert subspaces of H. Then $\bigcap_{n\ge1} G_n = G$ is a Hilbert subspace of H and, for all $x \in H$,

$$\lim_{n\uparrow\infty} P_{G_n}(x) = P_G(x).$$

Proof: (i) The set $\bigcup_{n\ge1} G_n$ is evidently a vector subspace of H (in general, however, it is not closed). Its closure, G, is a Hilbert subspace (Theorem C1.3). To any $y \in G$ one can associate a sequence $\{y_n\}_{n\ge1}$, where $y_n \in G_n$, and

$$\lim_{n\to\infty} \|y - y_n\| = 0.$$

Take $y = P_G(x)$. By the parallelogram identity,

$$\begin{aligned}\|P_{G_n}(x) - P_G(x)\|^2 &= \|(x - P_G(x)) - (x - P_{G_n}(x))\|^2 \\ &= 2\|x - P_{G_n}(x)\|^2 + 2\|x - P_G(x)\|^2 \\ &\quad - 4\|x - \tfrac{1}{2}(P_{G_n}(x) + P_G(x))\|.\end{aligned}$$

But since $P_{G_n}(x) + P_G(x)$ is a vector in G,

$$\|x - \tfrac{1}{2}(P_{G_n}(x) + P_G(x))\|^2 \geq \|x - P_G(x)\|^2,$$

and therefore,

$$\begin{aligned}\|P_{G_n}(x) - P_G(x)\|^2 &\leq 2\|x - P_{G_n}(x)\|^2 - 2\|x - P_G(x)\|^2 \\ &\leq 2\|x - y_n\|^2 - 2\|x - P_G(x)\|^2.\end{aligned}$$

By the continuity of the norm,

$$\lim_{n\uparrow\infty} \|x - y_n\|^2 = \|x - P_G(x)\|^2,$$

and, finally,

$$\lim_{n\uparrow\infty} \|P_{G_n}(x) - P_G(x)\|^2 = 0.$$

(ii) Devise a direct proof in the spirit of (i) or use the fact

$$G^{\perp} = \text{clos}\left(\bigcup_{n\geq 1} G_n^{\perp}\right). \qquad \blacksquare$$

EXERCISE **C1.6.** *Prove the following two assertions:*

- *Let $\{G_n\}$ be a nonincreasing sequence of Hilbert subspaces of H. Then $\bigcap_{n\geq 1} G_n = \varnothing$ if and only if $\lim_{n\uparrow\infty} P_{G_n}(x) = 0$ for all $x \in H$.*
- *Let $\{G_n\}_{n\geq 1}$ be a nondecreasing sequence of Hilbert subspaces of H. Then $\text{clos}\bigcup_{n\geq 1} G_n = H$ if and only if $\lim_{N\uparrow\infty} P_{G_n}(x) = x$ for all $x \in H$.*

NOTATION. *If G_1 and G_2 are orthogonal Hilbert subspaces of the Hilbert space H,*

$$G_1 \oplus G_2 := \{z = x_1 + x_2 : x_1 \in G,\ x_2 \in G_2\}$$

is called the orthogonal sum *of G_1 and G_2.*

Riesz's Representation Theorem

DEFINITION **C1.6.** *Let H be a Hilbert space over $\mathbb{C}$ and let $f : H \mapsto \mathbb{C}$ be a linear mapping; f is then called a (complex)* linear form *on H. It is said to be continuous if there exists $A \geq 0$ such that*

$$|f(x_1) - f(x_2)| \leq A\|x_1 - x_2\| \quad \textit{for all } x_1, x_2 \in H. \tag{12}$$

The infimum of the constants A satisfying (12) *is called the* norm *of f.*

EXAMPLE **C1.3.** *Let $y \in H$ and define $f : H \mapsto \mathbb{C}$ by*

$$f(x) = \langle x, y\rangle. \tag{13}$$

It is a linear form and, by Schwarz's inequality,

$$|f(x_1) - f(x_2)| = |f(x_1 - x_2)|$$
$$= |\langle x_1 - x_2, y\rangle| \le \|y\|\,\|x_1 - x_2\|.$$

Therefore, f is continuous. Its norm is $\|y\|$. To prove this it remains to show that if K is such that $|\langle x_1 - x_2, y\rangle| \le K\,\|x_1 - x_2\|$ for all $x_1, x_2 \in H$, then $\|y\| \le K$. It suffices to take $x_1 = x_2 = y$ above, which gives $\|y\|^2 \le K\|y\|$.

We now state and prove *Riesz's representation theorem.*

THEOREM **C1.9.** *Let $f : H \mapsto \mathbb{C}$ be a continuous linear form on the Hilbert space H. Then there exists a unique $y \in H$ such that* (13) *is true for all $x \in H$.*

Proof: Uniqueness. Let $y, y' \in H$ be such that

$$f(x) = \langle x, y\rangle = \langle x, y'\rangle \quad \text{for all } x \in H.$$

In particular,

$$\langle x, y - y'\rangle = 0 \quad \text{for all } x \in H.$$

The choice $x = y - y'$ leads to $\|y - y'\|^2 = 0$, that is, $y - y'$,

Existence: Consider the kernel of f, $N = \{u \in H : f(u) = 0\}$. It is a Hilbert subspace of H. We may suppose that f is not identically zero (otherwise, if $f \equiv 0$, take $y = 0$ in (13)). In particular, N is strictly included in H. This implies that $N^\perp$ does not reduces to the singleton $\{0\}$ and, therefore, there exists $z \in N^\perp$, $z \neq 0$. Define y by

$$y = f(z)^* \frac{z}{\|z\|^2}.$$

For all $x \in N$, $\langle x, y\rangle = 0$; therefore, in particular, $\langle x, y\rangle = f(x)$. Also,

$$\langle z, y\rangle = \left\langle z,\ f(z)^* \frac{z}{\|z\|^2}\right\rangle = f(z).$$

Therefore, the mappings $x \to f(x)$ and $x \to \langle x, y\rangle$ coincide on the Hilbert subspace generated by N and z. But this subspace is H itself. Indeed, for all $x \in H$,

$$x = \left(x - \frac{f(x)}{f(z)} z\right) + \frac{f(x)}{f(z)} z = u + w,$$

where $u \in N$ and w is colinear to z. ■

C2

Complete Orthonormal Systems

C2·1 Orthonormal Expansions

The result of this section is the pillar of the L^2-theory of Fourier series and wavelet expansions. It concerns the possibility of decomposing a vector of a Hilbert space along an orthonormal base.

The Gram–Schmidt Orthonormalization Procedure

The central notion is that of an orthonormal system:

DEFINITION **C2.1.** *The sequence* $\{e_n\}_{n\geq 0}$ *in a Hilbert space* H *is called an* orthonormal system *of* H *if it satisfies the following two conditions:*

(α) $\langle e_n, e_k \rangle = 0$ *for all* $n \neq k$; *and*
(β) $\|e_n\| = 1$ *for all* $n \geq 0$.

An orthonormal system $\{e_n\}_{n\geq 0}$ is *free* in the sense that an arbitrary *finite* subset of it is linearly independent. For example, taking $(e_1, \dots, e_k)$, the relation

$$\sum_{i=1}^{k} \alpha_i e_i = 0$$

implies that

$$\alpha_\ell = \left\langle e_\ell, \quad \sum_{i=1}^{k} \alpha_i e_i \right\rangle = 0 \quad 1 \leq \ell \leq k.$$

EXERCISE **C2.1.** *Let* $\{f_n\}_{n\geq 0}$ *be a sequence of vectors of a Hilbert space* H. *Construct* $\{e_n\}_{n\geq 0}$ *by the Gram–Schmidt orthonormalization procedure:*

- *Set $p(0) = 0$ and $e_0 = f_0/\|f_0\|$ (assuming $f_0 \neq 0$ without loss of generality);*
- *With $e_0, \dots, e_n$ and $p(n)$ defined, let $p(n+1)$ be the first index $p > p(n)$ such that f_p is independent of $e_0, \dots, e_n$, and define, with $p = p(n+1)$,*

$$e_{n+1} = \frac{f_p - \sum_{i=1}^{n} \langle f_p, e_i \rangle e_i}{\left\| f_p - \sum_{i=1}^{n} \langle f_p, e_i \rangle e_i \right\|}.$$

Show that $\{e_n\}_{n\geq 0}$ is an orthonormal system.

Hilbert Basis

The following theorem gives the preliminary results that we shall need for the proof of the Hilbert basis theorem.

THEOREM **C2.1.** *Let $\{e_n\}_{n\geq 0}$ be an orthonormal system of H and let G be the Hilbert subspace of H generated by $\{e_n\}_{n\geq 1}$. Then:*

(a) For an arbitrary sequence $\{\alpha_n\}_{n\geq 0}$ of complex numbers, the series $\sum_{n\geq 0} \alpha_n e_n$ is convergent in H if and only if $\{\alpha_n\}_{n\geq 1} \in \ell^2_{\mathbb{C}}$, in which case

$$\left\| \sum_{n\geq 0} \alpha_n e_n \right\|^2 = \sum_{n\geq 0} |\alpha_n|^2. \tag{14}$$

(b) For all $x \in H$, Bessel's inequality *holds:*

$$\sum_{n\geq 0} |\langle x, e_n \rangle|^2 \leq \|x\|^2. \tag{15}$$

(c) For all $x \in H$, the series $\sum_{n\geq 0} \langle x, e_n \rangle e_n$ converges, and

$$\sum_{n\geq 0} \langle x, e_n \rangle e_n = P_G(x), \tag{16}$$

where P_G is the projection on G.

(d) For all $x, y \in H$, the series $\sum_{n\geq 1} \langle x, e_n \rangle \langle y, e_n \rangle$ is absolutely convergent, and

$$\sum_{n\geq 0} \langle x, e_n \rangle \langle y, e_n \rangle^* = \langle P_G(x), P_G(y) \rangle. \tag{17}$$

Proof: (a) From Pythagoras' theorem we have

$$\left\| \sum_{j=m+1}^{n} \alpha_j e_j \right\|^2 = \sum_{j=m+1}^{n} |\alpha_j|^2,$$

and, therefore, $\{\sum_{j=0}^{n} \alpha_j e_j\}_{n\geq 0}$ is a Cauchy sequence in H if and only if $\{\sum_{j=0}^{n} |\alpha_j|^2\}_{n\geq 0}$ is a Cauchy sequence in $\mathbb{R}$. In other words, $\sum_{n\geq 0} \alpha_n e_n$ converges if and only if $\sum_{n\geq 0} |\alpha_n|^2 < \infty$. In this case equality (14) follows from the continuity of the norm, by letting n tend to ∞ in the last display.

(b) According to (α) of Theorem C1.7, $\|x\| \geq \|P_{G_n}(x)\|$, where G_n is the Hilbert subspace spanned by $\{e_1, \ldots, e_n\}$. But

$$P_{G_n}(x) = \sum_{i=0}^{n} \langle x, e_i\rangle e_i,$$

and by Pythagoras' theorem,

$$\|P_{G_n}(x)\|^2 = \sum_{i=0}^{n} |\langle x, e_i\rangle|^2.$$

Therefore,

$$\|x\|^2 \geq \sum_{i=0}^{n} |\langle x, e_i\rangle|^2,$$

from which Bessel's inequality follows on letting $n \to \infty$.

(c) From (15) and result (a), it follows that the series $\sum_{n\geq 0} \langle x, e_n\rangle e_n$ converges. For any $m \geq 0$ and for all $N \geq m$,

$$\left\langle x - \sum_{n=0}^{N} \langle x, e_n\rangle e_n, e_m \right\rangle = 0,$$

and, therefore, by continuity of the Hermitian product,

$$\left\langle x - \sum_{n\geq 0} \langle x, e_n\rangle e_n, e_m \right\rangle = 0 \quad \text{for all } m \geq 0.$$

This implies that $x - \sum_{n\geq 0} \langle x, e_n\rangle e_n$ is orthogonal to G. Also, $\sum_{n\geq 0} \langle x, e_n\rangle e_n \in G$. Therefore, by the projection principle,

$$P_G(x) = \sum_{n\geq 0} \langle x, e_n\rangle e_n.$$

(d) By Schwarz's inequality in $\ell^2_{\mathbb{C}}$, for all $N \geq 0$,

$$\left(\sum_{n=0}^{N} |\langle x, e_n\rangle \langle y, e_n\rangle^*|\right)^2 \leq \left(\sum_{n=0}^{N} |\langle x, e_n\rangle|^2\right)\left(\sum_{n=0}^{N} |\langle y, e_n\rangle|^2\right)$$
$$\leq \|x\|^2 \, \|y\|^2.$$

Therefore, the series $\sum_{n=0}^{\infty} \langle x, e_n\rangle \langle y, e_n\rangle^*$ is absolutely convergent. Also, by an elementary computation,

$$\left\langle \sum_{n=0}^{N} \langle x, e_n\rangle e_n, \sum_{n=0}^{N} \langle y, e_n\rangle e_n \right\rangle = \sum_{n=0}^{N} \langle x, e_n\rangle \langle y, e_n\rangle^*.$$

Letting $N \to \infty$, we obtain (17) (using (16) and the continuity of the Hermitian product). ■

DEFINITION **C2.2.** *The sequence $\{w_n\}_{n\geq 0}$ of vectors of H is said to be total in H if it generates H.*

In other words, the finite linear combination of the elements of $\{w_n\}_{n\geq 0}$ forms a dense subset of H.

EXERCISE **C2.2.** *Prove that a sequence $\{w_n\}_{n\geq 0}$ of the Hilbert space H is total in H if and only if there is no element of H orthogonal to all the w_n, $n \geq 0$, except 0, that is, if and only if*

$$(\langle z, w_n\rangle = 0 \text{ for all } n \geq 0) \implies (z = 0). \tag{18}$$

We are now ready for the fundamental result: the *Hilbert basis theorem.*

THEOREM **C2.2.** *Let $\{e_n\}_{n\geq 0}$ be an orthonormal system of H. The following properties are equivalent:*

(a) $\{e_n\}_{n\geq 0}$ is total in H;
(b) for all $x \in H$, the Plancherel–Parseval identity holds true:

$$\|x\|^2 = \sum_{n\geq 0} |\langle x, e_n\rangle|^2; \tag{19}$$

(c) for all $x \in H$,

$$x = \sum_{n\geq 0} \langle x, e_n\rangle e_n. \tag{20}$$

Proof: (a)⇒(c) According to (c) of Theorem C2.1,

$$\sum_{n\geq 0} \langle x, e_n\rangle e_n = P_G(x),$$

where G is the Hilbert subspace generated by $\{e_n\}_{n\geq 0}$. Since $\{e_n\}_{n\geq 0}$ is total, it follows by (18) that $G^\perp = \{0\}$, and therefore $P_G(x) = x$.

(c)⇒(b) This follows from (a) of Theorem C2.1.

(b)⇒(a) From (14) and (16),

$$\sum_{n\geq 0} |\langle x, e_n\rangle|^2 = \|P_G(x)\|^2,$$

and (19) therefore implies

$$\|x\|^2 = \|P_G(x)\|^2.$$

From Pythagoras' theorem,

$$\begin{aligned}\|x\|^2 &= \|P_G(x) + x - P_G(x)\|^2 \\ &= \|P_G(x)\|^2 + \|x - P_G(x)\|^2 \\ &= \|x\|^2 + \|x - P_G(x)\|^2;\end{aligned}$$

therefore,

$$\|x - P_G(x)\|^2 = 0,$$

which implies

$$x = P_G(x).$$

Since this is true for all $x \in H$, we must have $G \equiv H$, that is, $\{e_n\}_{n\geq 0}$ is total in H. ■

A sequence $\{e_n\}_{n\geq 0}$ satisfying one (and then all) of the conditions of Theorem C2.2 is called a (denumerable) *Hilbert basis* of H.

EXERCISE **C2.3.** *Let ψ be a function in $L^2_{\mathbb{R}}(\mathbb{R})$ with the FT $\widehat{\psi} = \frac{1}{2\pi}1_I$, where $I = [-2\pi, -\pi]\cup[+\pi, +2\pi]$. Show that $\{\psi_{j,n}\}_{j\in\mathbb{Z},n\in\mathbb{Z}}$ is a Hilbert basis of $L^2_{\mathbb{R}}(\mathbb{R})$, where $\psi_{j,n}(x) = 2^{j/2}\psi(2^{j/2}x - n)$.*

EXERCISE **C2.4.** *Let $\{g_j\}_{j\geq 0}$ be a Hilbert basis of $L^2((0, 1])$. Show that $\{g_j(\cdot - n)1_{(n,n+1]}(\cdot)\}_{j\geq 0,n\in\mathbb{Z}}$ is a Hilbert basis of $L^2(\mathbb{R})$. (Here, $L^2(I)$ Denotes the Hilbert space (equivalence classes) of measurable complex-valued functions defined on I, with the Hermitian product $\langle f, g\rangle = \int_I f(t)g(t)^*\mathrm{d}t$.)*

Biorthonormal Expansions

DEFINITION **C2.3.** *Two sequences $\{e_n\}_{n\geq 0}$ and $\{d_n\}_{n\geq 0}$ of a Hilbert space H form a* biorthonormal system *if*

(α) $\langle e_n, d_k\rangle = 0$ for all $n \neq k$,
(β) $\langle e_n, d_n\rangle = 1$ for all $n \geq 0$.

This system is called complete *if, in addition, each of the sequences $\{e_n\}_{n\geq 0}$ and $\{d_n\}_{n\geq 0}$ forms a total subset of H.*

Then we have the biorthonormal expansions

$$x = \sum_{n\geq 0}\langle x, e_n\rangle d_n, \qquad x = \sum_{n\geq 0}\langle x, d_n\rangle e_n$$

whenever these series converge. Indeed, with the first series, for example, calling its sum y, we have for any integer $m \geq 0$,

$$\langle y, e_m\rangle = \left\langle \sum_{n\geq 0}\langle x, e_n\rangle d_n, e_m\right\rangle$$

$$= \sum_{n\geq 0}\langle x, e_n\rangle\langle d_n, e_m\rangle = \langle x, e_m\rangle.$$

Therefore,

$$\langle x - y, e_m\rangle = 0 \quad \text{for all } m \geq 0.$$

Since $\{e_n\}_{n\geq 0}$ is total in H, this implies $x - y = 0$.

Separable Hilbert Spaces

An interesting theoretical question is: For what type of Hilbert spaces is there a denumerable Hilbert basis? Here is a first (theoretical) answer.

DEFINITION **C2.4.** *A Hilbert space H is called a* separable Hilbert space *if it contains a sequence $\{f_n\}_{n\geq 0}$ that is dense in H.*

THEOREM **C2.3.** *A separable Hilbert space admits at least one denumerable Hilbert basis.*

Proof: Let $\{f_n\}_{n\geq 0}$ be a sequence defined in Definition C2.4. Construct from it the orthonormal sequence $\{e_n\}_{n\geq 0}$ by the *Gram–Schmidt* orthonormalization procedure. It is a Hilbert basis because (a) of Theorem C2.2 is satisfied. Indeed, for any $z \in H$,

$$(\langle e_n, z\rangle = 0 \text{ for all } n \geq 0) \implies (\langle f_p, z\rangle = 0 \text{ for all} n \geq 0).$$

In particular, $\langle y, z\rangle = 0$ for any finite linear combination of $\{f_p\}_{p\geq 0}$. Because $\{f_p\}_{p\geq 0}$ is dense, $\langle y, z\rangle = 0$ for all $y \in H$. In particular, $\langle z, z\rangle = 0$, that is, $z = 0$. ■

C2·2 Two Important Hilbert Bases

The Fourier Basis

The following theorem is the fundamental result of the theory of Fourier series of finite-power periodic signals.

THEOREM **C2.4.** *The sequence*

$$\{e_n(\cdot)\} \stackrel{\text{def}}{=} \left\{\frac{1}{\sqrt{T}} e^{2i\pi \frac{n}{T}\cdot}\right\}, \quad n \in \mathbb{Z},$$

is a Hilbert basis of $L^2_{\mathbb{C}}([0, T])$.

Proof: One first observes that $\{e_n(\cdot),\ n \in \mathbb{Z}\}$ is an orthonormal system in $L^2_{\mathbb{C}}([0, T])$. It remains to show that the linear space it generates is dense in $L^2_{\mathbb{C}}([0, T])$ (Theorem C2.2).

For this, let $f(t) \in L^2_{\mathbb{C}}([0, T])$ and let $f_N(t)$ be its projection on the Hilbert subspace generated by $\{e_n(\cdot),\ -N \leq n \leq N\}$. The coefficient of e_n in this projection is $c_n(f) = \langle f, e_n\rangle_{L^2_{\mathbb{C}}([0,T])}$, we have

$$\sum_{n=-N}^{+N} |c_n(f)|^2 + \int_0^T |f(t) - f_N(t)|^2\, dt = \int_0^T |f(t)|^2\, dt. \tag{21}$$

(This is Pythagoras' theorem for projections: $\|P_G(x)\|^2 + \|x - P_G(x)\|^2 = \|x\|^2$.) In particular, $\sum_{n\in\mathbb{Z}} |c_n(f)|^2 < \infty$. It remains to show ((b) of Theorem C2.2) that

$$\lim_{N\uparrow\infty} \int_0^T |f(t) - f_N(t)|^2 \, dt = 0.$$

We assume in a first step that f is continuous. For such a function, the formula

$$\varphi(x) = \int_0^T \tilde{f}(x+t)\tilde{f}(t)^* \, dt,$$

where

$$\tilde{f}(t) = \sum_{n\in\mathbb{Z}} f(t+nT)1_{(0,T]}(t+nT),$$

defines a T-periodic and continuous function φ. Its nth Fourier coefficient is

$$\begin{aligned} c_n(\varphi) &= \frac{1}{T}\int_0^T \left(\tilde{f}(x+t)\tilde{f}(t)^* \, dt\right) e^{-2i\pi\frac{n}{T}x} \, dx, \\ &= \frac{1}{T}\int_0^T \tilde{f}(t)^* \left\{\int_0^T \tilde{f}(x+t)e^{-2i\pi\frac{n}{T}x} \, dx\right\} dt \\ &= \frac{1}{T}\int_0^T f(t)^* \left\{\int_t^{t+T} \tilde{f}(s)e^{-2i\pi\frac{n}{T}s} \, ds\right\} e^{2i\pi\frac{n}{T}t} \, dt \\ &= T|c_n(f)|^2. \end{aligned}$$

Since $\sum_{n\in\mathbb{Z}} |c_n(f)|^2 < \infty$ and $\varphi(x)$ is continuous, it follows from the Fourier inversion theorem for locally integrable periodic functions that, for all $x \in \mathbb{R}$,

$$\varphi(x) = \sum_{n\in\mathbb{Z}} |c_n(f)|^2 e^{2i\pi\frac{n}{T}x}.$$

In particular, for $x = 0$,

$$\varphi(0) = \int_0^T |f(t)|^2 \, dt = \sum_{n\in\mathbb{Z}} |c_n(f)|^2,$$

and therefore, in view of (21),

$$\lim_{N\uparrow\infty} \int_0^T |f(t) - f_N(t)|^2 \, dt = 0.$$

It remains to pass from the continuous functions to the square-integrable functions. Since the space $C([0,T])$ of continuous functions from $[0,T]$ into $\mathbb{C}$ is dense in $L^2_{\mathbb{C}}([0,T])$ (Theorem 27), with any $\varepsilon > 0$, one can associate $\varphi \in C([0,T])$ such that $\|f - \varphi\| \le \varepsilon/3$. By Bessel's inequality, $\|f_N - \varphi_N\|^2 = \|(f - \varphi)_N\|^2 \le$

$\|f - \varphi\|^2$, and therefore,

$$\begin{aligned}\|f - f_N\| &\leq \|f - \varphi\| + \|\varphi - \varphi_N\| + \|f_N - \varphi_N\| \\ &\leq \|\varphi - \varphi_N\| + 2\|f - \varphi\| \\ &\leq \|\varphi - \varphi_N\| + 2\frac{\varepsilon}{3}.\end{aligned}$$

For N sufficiently large, $\|\varphi - \varphi_N\| \leq \varepsilon/3$. Therefore, for N sufficiently large, $\|f - f_N\| \leq \varepsilon$. ■

The Cardinal Sine Basis

The L^1-version of the Shannon–Nyquist theorem of Section B2·1 contains a condition bearing on the samples themselves, namely,

$$\sum_{n \in \mathbb{Z}} \left| s\left(\frac{n}{2B}\right) \right| < \infty. \tag{22}$$

The simplest way of removing this unaesthetic condition is given by the *L^2-version of the Shannon–Nyquist theorem.*

THEOREM **C2.5.** *Let $s(t)$ be a base-band (B) signal of finite energy. Then*

$$\lim_{N \uparrow \infty} \int_{\mathbb{R}} \left| s(t) - \sum_{n=-N}^{+N} b_n \operatorname{sinc}(2Bt - n) \right|^2 dt = 0, \tag{23}$$

where

$$b_n = \int_{-B}^{+B} \hat{s}(\nu)\, e^{2i\pi\nu \frac{n}{2B}}\, d\nu.$$ ■

Proof: Let $L^2_{\mathbb{C}}(\mathbb{R}; B)$ be the Hilbert subspace of $L^2_{\mathbb{C}}(\mathbb{R})$ consisting of the finite-energy complex signals with a Fourier transform having its support contained in $[-B, +B]$. The sequence

$$\left\{ \frac{1}{\sqrt{2B}} h\left(\cdot - \frac{n}{2B}\right) \right\}_{n \in \mathbb{Z}}, \tag{24}$$

where $h(t) \equiv 2B \operatorname{sinc}(2Bt)$, is an orthonormal basis of $L^2_{\mathbb{C}}(\mathbb{R}; B)$. Indeed, the functions of this system are in $L^2_{\mathbb{C}}(\mathbb{R}; B)$, and they form an orthonormal system since, by the Plancherel–Parseval formula,

$$\begin{aligned}\int_{\mathbb{R}} h\left(t - \frac{n}{2B}\right) h\left(t - \frac{k}{2B}\right) dt &= \int_{\mathbb{R}} \hat{h}(\nu) e^{-2i\pi\nu \frac{n}{2B}} \left(\hat{h}(\nu) e^{-2i\pi\nu \frac{k}{2B}}\right)^* d\nu \\ &= \int_{-B}^{+B} e^{2i\pi\nu \frac{k-n}{2B}}\, d\nu = 2B \times 1_{n=k}.\end{aligned}$$

It remains to prove the totality of the orthonormal system (24) (see Theorem C2.2). We must show that if $g(t) \in L^2_{\mathbb{C}}(\mathbb{R}; B)$ and

$$\int_{\mathbb{R}} g(t)h\left(t - \frac{n}{2B}\right) dt = 0 \quad \text{for all } n \in \mathbb{Z}, \tag{25}$$

then $g(t) \equiv 0$ as a function of $L^2_{\mathbb{C}}(\mathbb{R}; B)$ (or, equivalently, that $g(t) = 0$ almost everywhere).

Condition (25) is equivalent (by the Plancherel–Parseval identity) to

$$\int_{-B}^{+B} \hat{g}(\nu)e^{2i\pi\nu\frac{n}{2B}}\, d\nu = 0 \quad \text{for all } n \in \mathbb{Z}. \tag{26}$$

But we have proven in the previous section that the system $\{e^{2i\pi\nu n/2B}\}_{n\in\mathbb{Z}}$ is total in $L^2_{\mathbb{C}}(\mathbb{R}; B)$; therefore, (26) implies $\hat{g}(\nu) = 0$ almost everywhere, and consequently, $g(t) = 0$ almost everywhere.

Expanding $s(t) \in L^2_{\mathbb{C}}(\mathbb{R}; B)$ in the Hilbert basis (24) yields

$$s(t) = \lim_{N\uparrow\infty} \sum_{-N}^{+N} c_n \frac{1}{\sqrt{2B}} h\left(t - \frac{n}{2B}\right), \tag{27}$$

where the limit and the equality in (27) are taken in the L^2-sense (as in (23)), and

$$c_n = \int_{\mathbb{R}} s(t) \frac{1}{\sqrt{2B}} h\left(t - \frac{n}{2B}\right) dt.$$

By the Plancherel–Parseval identity,

$$c_n = \int_{-B}^{+B} \hat{s}(\nu) \frac{1}{\sqrt{2B}} e^{2i\pi\nu\frac{n}{2B}}\, d\nu. \qquad \blacksquare$$

An Apparent Paradox

Note that since $\hat{s}(\nu)$ is in L^2 and of compact support, it is also in L^1, and therefore the Fourier inversion formula is true and the reconstruction formula takes the familiar form

$$s(t) = \sum_{n\in\mathbb{Z}} s\left(\frac{n}{2B}\right) \operatorname{sinc}(2Bt - n). \tag{28}$$

This is essentially true, but not quite.

Indeed, imagine that someone tells you the following: Look, I have an proof that a L^2-signal $s(t)$, base-band (B) is almost everywhere zero! Here is my cute proof. Of course, the reconstruction equality is in the sense of equality of L^2-functions, and in particular, it holds only for almost all t. Now, let me change the original signal to obtain a new signal $s'(t)$ differing from $s(t)$ only at the times $n/2B$, where I set $s'(n/2B) = 0$. I now apply the reconstruction formula and obtain that $s'(t)$ is almost everywhere zero. But $s'(t)$ and $s(t)$ are almost everywhere equal. Therefore, $s(t)$ is almost everywhere zero! Quod erat demonstrandum.

The flaw in the above "proof" is that the Fourier inversion formula holds only almost everywhere, and maybe not at the sampling times. Therefore, formula (28) is true only if the Fourier inversion formula can be applied at all the times of the form $n/2B$. This is the case if $s(t)$ is continuous, because the inversion formula then holds *everywhere*.

We see that the continuity hypothesis always pops up. We cannot expect a much better version of the sampling theorem in the L^1 or L^2 framework. Indeed, since $\hat{s}(\nu)$ is integrable, the right-hand side of

$$s(t) = \int_{\mathbb{R}} \hat{s}(\nu) e^{2i\pi\nu t} d\nu$$

is continuous, and $s(t)$ is therefore almost everywhere continuous.

We have a sampling theorem for sinusoids and for decomposable signals (Theorem B2.5), and those signals are neither in L^1 nor in L^2. Note, however, that they are continuous.

C3

Fourier Transforms of Finite Energy Signals

C3·1 Fourier Transform in L^2

A stable signal as simple as the rectangular pulse has a Fourier transform that is not integrable, and therefore one cannot use the Fourier inversion theorem for stable signals as it is. However, there is a version of this inversion formula that applies to all finite-energy functions (for instance, the rectangular pulse). The analysis becomes slightly more involved, and we will have to use the framework of Hilbert spaces. This is largely compensated by the formal beauty of the results, due to the fact that a square-integrable function and its FT play symmetrical roles.

The Isometric Extension

We start with a technical result. We use $f(.)$ to denote the function $f : \mathbb{R} \mapsto \mathbb{C}$; in particular, $f(a + .)$ is the function $f_a : \mathbb{R} \mapsto \mathbb{C}$ defined by $f_a(t) = f(a + t)$.

THEOREM **C3.1.** *Let $s(t) \in L^2_{\mathbb{C}}(\mathbb{R})$. The mapping from $\mathbb{R}$ into $L^2_{\mathbb{C}}(\mathbb{R})$ defined by*

$$t \to s(t + \cdot)$$

is uniformly continuous.

Proof: We have to prove that the quantity

$$\int_{\mathbb{R}} |s(t + h + u) - s(t + u)|^2 \, du = \int_{\mathbb{R}} |s(h + u) - s(u)|^2 \, du$$

tends to 0 when $h \to 0$. When $s(\cdot)$ is continuous and compactly supported, the result follows by dominated convergence. The general case where $s(\cdot) \in L^2_{\mathbb{C}}(\mathbb{R})$

is obtained by approximating $s(\cdot)$ in $L^2_{\mathbb{C}}(\mathbb{R})$ by continuous compactly supported functions (see the proof of Theorem A1.4). ■

From Schwarz's inequality, we deduce that

$$t \to \langle s(t+\cdot), s(\cdot)\rangle_{L^2_{\mathbb{C}}(\mathbb{R})}$$

is uniformly continuous on $\mathbb{R}$ and bounded by the energy of the signal.

The above function is

$$t \to \int_{\mathbb{R}} s(t+x)s^*(x)\,\mathrm{d}x \tag{29}$$

and called the *autocorrelation function* of the finite-energy signal $s(t)$. Note that it is the convolution $s(t) * \tilde{s}(t)$, where $\tilde{s}(t) = s(-t)^*$.

THEOREM **C3.2.** *If the complex signal $s(t)$ lies in $L^1_{\mathbb{C}}(\mathbb{R}) \cap L^2_{\mathbb{C}}(\mathbb{R})$, then its FT $\hat{s}(\nu)$ belongs to $L^2_{\mathbb{C}}(\mathbb{R})$ and*

$$\int_{\mathbb{R}} |s(t)|^2\,\mathrm{d}t = \int_{\mathbb{R}} |\hat{s}(\nu)|^2\,\mathrm{d}\nu. \tag{30}$$

Proof: The signal $\tilde{s}(t)$ admits $\hat{s}(\nu)^*$ as FT, and thus by the convolution–multiplication rule,

$$(s * \tilde{s})(t) \overset{\mathrm{FT}}{\to} |\hat{s}(\nu)|^2. \tag{31}$$

Consider the Gaussian density function

$$h_\sigma(t) = \frac{1}{\sigma\sqrt{2\pi}}\, e^{-\frac{t^2}{2\sigma^2}}.$$

Applying the result in (14) of Chapter A1, with $(s * \tilde{s})(t)$ instead of $s(t)$, and observing that $h_\sigma(t)$ is an even function, we obtain

$$\int_{\mathbb{R}} |\hat{s}(\nu)|^2 \hat{h}_\sigma(\nu)\,\mathrm{d}\nu = \int_{\mathbb{R}} (s * \tilde{s})(x) h_\sigma(x)\,\mathrm{d}x. \tag{32}$$

Since $\hat{h}_\sigma(\nu) = e^{-2\pi^2\sigma^2x^2} \uparrow 1$ when $\sigma \downarrow 0$, the left-hand side of (32) tends to $\int_{\mathbb{R}} |\hat{s}(\nu)|^2\,\mathrm{d}\nu$, by dominated convergence.

On the other hand, since the autocorrelation function $(s * \tilde{s})(t)$ is continuous and bounded, the quantity

$$\int_{\mathbb{R}} (s * \tilde{s})(x) h_\sigma(x)\,\mathrm{d}x = \int_{\mathbb{R}} (s * \tilde{s})(\sigma y) h_1(y)\,\mathrm{d}y$$

tends when $\sigma \downarrow 0$ toward

$$\int_{\mathbb{R}} (s * \tilde{s})(0) h_1(y)\,\mathrm{d}y = (s * \tilde{s})(0) = \int_{\mathbb{R}} |s(t)|^2\,\mathrm{d}t,$$

by dominated convergence. ■

From the last theorem, we have that the mapping $\varphi : s(t) \to \hat{s}(\nu)$ from $L^1_{\mathbb{C}}(\mathbb{R}) \cap L^2_{\mathbb{C}}(\mathbb{R})$ into $L^2_{\mathbb{C}}(\mathbb{R})$ thus defined is isometric and linear. Since $L^1 \cap L^2$ is dense in L^2, this linear isometry can be uniquely extended into a linear isometry from $L^2_{\mathbb{C}}(\mathbb{R})$ into itself (Theorem C1.5). We will continue both to denote by $\hat{s}(\nu)$ the image of $s(t)$ under this isometry and to call it the FT of $s(t)$.

EXERCISE **C3.1.** *Show that for* $s(t) \in L^2_{\mathbb{C}}(\mathbb{R})$,

$$\lim_{T \uparrow \infty} \int_{\mathbb{R}} \left| \hat{s}(\nu) - \int_{-T}^{+T} s(t) e^{-2i\pi\nu t} \mathrm{d}t \right|^2 \mathrm{d}\nu. \tag{33}$$

The above isometry is expressed by the *Plancherel–Parseval identity*:

THEOREM **C3.3.** *If* $s_1(t)$ *and* $s_2(t)$ *are finite-energy, complex signals, then*

$$\int_{\mathbb{R}} s_1(t) s_2(t)^* \, \mathrm{d}t = \int_{\mathbb{R}} \hat{s}_1(\nu) \hat{s}_2(\nu)^* \, \mathrm{d}\nu. \tag{34}$$

EXERCISE **C3.2.** *Show that*

$$\int_{\mathbb{R}} \left(\frac{\sin(\pi\nu)}{\pi\nu} \right)^2 \mathrm{d}\nu = 1.$$

THEOREM **C3.4.** *If* $h(t) \in L^1_{\mathbb{C}}(\mathbb{R})$ *and* $x(t) \in L^2_{\mathbb{C}}(\mathbb{R})$, *then*

$$y(t) = \int_{\mathbb{R}} h(t-s) x(s) \, \mathrm{d}s \tag{35}$$

is almost everywhere well defined and in $L^2_{\mathbb{C}}(\mathbb{R})$. *Furthermore, its FT is*

$$\hat{y}(\nu) = \hat{h}(\nu) \hat{x}(\nu). \tag{36}$$

Proof: Let us first show that $\int_{\mathbb{R}} h(t-s) x(s) \, \mathrm{d}s$ is well defined. For this we observe that on the one hand

$$\int_{\mathbb{R}} |h(t-s)| \, |x(s)| \, \mathrm{d}s \le \int_{\mathbb{R}} |h(t-s)| (1 + |x(s)|^2) \, \mathrm{d}s$$

$$= \int_{\mathbb{R}} |h(t)| \, \mathrm{d}t + \int_{\mathbb{R}} |h(t-s)| \, |x(s)|^2 \, \mathrm{d}s,$$

and on the other, for almost all t,

$$\int_{\mathbb{R}} |h(t-s)| \, |(x(s)|^2 \, \mathrm{d}s < \infty,$$

since $|h(t)|$ and $|x(t)|^2$ are in $L^1_{\mathbb{C}}(\mathbb{R})$. Therefore, for almost all t,

$$\int_{\mathbb{R}} |h(t-s| \, |x(s)| \, \mathrm{d}s < \infty,$$

and $y(t)$ is almost everywhere well defined. We now show that $y(t) \in L^2_{\mathbb{C}}(\mathbb{R})$.

Using Fubini's theorem and Schwarz's inequality, we have

$$\begin{aligned}
&\int_{\mathbb{R}}\left|\int_{\mathbb{R}} h(t-s)x(s)\,\mathrm{d}s\right|^2 \mathrm{d}t \\
&= \int_{\mathbb{R}}\left|\int_{\mathbb{R}} h(u)x(t-u)\,\mathrm{d}u\right|^2 \mathrm{d}t \\
&= \int_{\mathbb{R}}\int_{\mathbb{R}}\left\{\int_{\mathbb{R}} x(t-u)x(t-v)^*\,\mathrm{d}t\right\} h(u)h(v)^*\,\mathrm{d}u\,\mathrm{d}v \\
&\le \left(\int_{\mathbb{R}} |x(s)|^2\,\mathrm{d}s\right)\left(\int_{\mathbb{R}} |h(u)|\,\mathrm{d}u\right)^2 < \infty.
\end{aligned}$$

For future reference, we rewrite this as

$$\|h * x\|_{L^2_{\mathbb{C}}(\mathbb{R})} \le \|h\|_{L^1_{\mathbb{C}}(\mathbb{R})}\|x\|_{L^2_{\mathbb{C}}(\mathbb{R})}. \tag{37}$$

The signal (35) is thus in $L^2_{\mathbb{C}}(\mathbb{R})$ when $h(t) \in L^1_{\mathbb{C}}(\mathbb{R})$ and $x(t) \in L^2_{\mathbb{C}}(\mathbb{R})$. If, furthermore, $x(t)$ is in $L^1_{\mathbb{C}}(\mathbb{R})$, then $y(t)$ is in $L^1_{\mathbb{C}}(\mathbb{R})$. Therefore,

$$x(t) \in L^1_{\mathbb{C}}(\mathbb{R}) \cap L^2_{\mathbb{C}}(\mathbb{R}) \to y(t) \in L^1_{\mathbb{C}}(\mathbb{R}) \cap L^2_{\mathbb{C}}(\mathbb{R}). \tag{38}$$

In this case, we obtain (36) by the convolution–multiplication formula in L^1.

We now suppose that $x(t)$ is in $L^2_{\mathbb{C}}(\mathbb{R})$ (but not necessarily in $L^1_{\mathbb{C}}(\mathbb{R})$). The signal

$$x_A(t) = x(t)1_{[-A,+A]}(t)$$

is in $L^1_{\mathbb{C}}(\mathbb{R}) \cap L^2_{\mathbb{C}}(\mathbb{R})$ and $\lim x_A(t) = x(t)$ in $L^2_{\mathbb{C}}(\mathbb{R})$. In particular, $\lim \hat{x}_A(\nu) = \hat{x}(\nu)$ in $L^2_{\mathbb{C}}(\mathbb{R})$. Introducing

$$y_A(t) = \int_{\mathbb{R}} h(t-s)x_A(s)\,\mathrm{d}s,$$

we have $\hat{y}_A(\nu) = \hat{h}(\nu)\hat{x}_A(\nu)$. Also, $\lim y_A(t) = y(t)$ in $L^2_{\mathbb{C}}(\mathbb{R})$ [use (37)], and thus $\lim \hat{y}_A(\nu) = \hat{y}(\nu)$ in $L^2_{\mathbb{C}}(\mathbb{R})$. Now, since $\lim \hat{x}_A(\nu) = \hat{x}(\nu)$ in $L^2_{\mathbb{C}}(\mathbb{R})$ and $\hat{h}(\nu)$ is bounded, $\lim \hat{h}(\nu)\hat{x}_A(\nu) = \hat{h}(\nu)\hat{x}(\nu)$ in $L^2_{\mathbb{C}}(\mathbb{R})$. Therefore, we have (36). ■

EXERCISE **C3.3.** *Use the Plancherel–Parseval identity to prove that*

$$\int_{\mathbb{R}} \frac{\mathrm{d}t}{(t^2+a^2)(t^2+b^2)} = \frac{\pi}{ab(a+b)}.$$

EXERCISE **C3.4.** *Show that*

$$H = \{w(t) \in L^2_{\mathbb{C}}(\mathbb{R});\, tw(t) \in L^2_{\mathbb{C}}(\mathbb{R}),\, \nu\widehat{w}(\nu) \in L^2_{\mathbb{C}}(\mathbb{R})\}$$

is a Hilbert space when endowed with the norm

$$\|w\|_H = \left(\|w\|^2_{L^2} + \|tw\|^2_{L^2} + \|\nu\widehat{w}\|^2_{L^2}\right)^{\frac{1}{2}}.$$

Show that the subset of H consisting of the C^∞-functions with compact support is dense in H. Hint: Select any φ in a C^∞-function with compact support, with *integral equal to* 1, *and equal to* 1 *in a neighborhood of* 0, *and for any* $w \in H$, *consider the function* $w(t)\varphi(t/n) * n\varphi(nt)$.

C3·2 Inversion Formula in L^2

So far, we know that the mapping $\varphi : L^2_{\mathbb{C}}(\mathbb{R}) \mapsto L^2_{\mathbb{C}}(\mathbb{R})$ defined in Section C3·1 is linear, isometric, and *into*. We shall now show that it is *onto*, and therefore bijective.

THEOREM **C3.5.** *Let* $\hat{s}(\nu)$ *be the FT of* $s(t) \in L^2_{\mathbb{C}}(\mathbb{R})$. *Then*

$$\varphi : \hat{s}(-\nu) \to s(t), \tag{39}$$

that is,

$$s(t) = \lim_{A\uparrow\infty} \int_{-A}^{+A} \hat{s}(\nu)e^{2i\pi\nu t}\,\mathrm{d}\nu, \tag{40}$$

where the limit is in $L^2_{\mathbb{C}}(\mathbb{R})$, *and the equality is almost everywhere.*

We shall prepare the way for the proof with the following result.

LEMMA **C3.1.** *Let* $u(t)$ *and* $v(t)$ *be two finite-energy signals. Then*

$$\int_{\mathbb{R}} u(x)\hat{v}(x)\,\mathrm{d}x = \int_{\mathbb{R}} \hat{u}(x)v(x)\,\mathrm{d}x. \tag{41}$$

Proof: If (41) is true for $u(t), v(t) \in L^1_{\mathbb{C}}(\mathbb{R}) \cap L^2_{\mathbb{C}}(\mathbb{R})$, then it also holds for $u(t), v(t) \in L^2_{\mathbb{C}}(\mathbb{R})$. Indeed, denoting $x_A(t) = x(t)1_{[-A,+A]}(t)$, we have

$$\int_{\mathbb{R}} u_A(x)\widehat{(v_A)}(x)\,\mathrm{d}x = \int_{\mathbb{R}} \widehat{(u_A)}(x)v_A(x)\,\mathrm{d}x,$$

that is, $\langle u_A, \widehat{v_A}\rangle = \langle \widehat{u_A}, v_A\rangle$. Now u_A, v_A, $\widehat{u_A}$, and $\widehat{v_A}$ tend in $L^2_{\mathbb{C}}(\mathbb{R})$ to $u, v, \hat{u}$, and $\hat{v}$, respectively, as $A \uparrow \infty$, and therefore, by the continuity of the Hermitian product, $\langle u, \hat{v}\rangle = \langle \hat{u}, v\rangle$.

The proof of (41) for stable signals is accomplished by Fubini's theorem:

$$\begin{aligned}\int_{\mathbb{R}} u(x)\hat{v}(x)\,\mathrm{d}x &= \int_{\mathbb{R}} u(x)\left\{\int_{\mathbb{R}} v(y)e^{-2i\pi xy}\,\mathrm{d}y\right\}\mathrm{d}x\\ &= \int_{\mathbb{R}} v(y)\left\{\int_{\mathbb{R}} u(x)e^{-2i\pi xy}\,\mathrm{d}y\right\}\mathrm{d}y\\ &= \int_{\mathbb{R}} v(y)\hat{u}(y)\,\mathrm{d}y.\end{aligned}$$

■

Proof of (39): Let $g(t)$ be a real signal in $L^2_{\mathbb{C}}(\mathbb{R})$, and define $f(t) = \widehat{(g^-)}(t)$, where $g^-(t) = g(-t)$. We have $\hat{f}(\nu) = \hat{g}(\nu)^*$. Therefore, by (41):

$$\int_{\mathbb{R}} g(x)\hat{f}(x)\,dx = \int_{\mathbb{R}} \hat{g}(x) f(x)\,dx$$
$$= \int \hat{g}(x)\hat{g}(x)^*\,dx.$$

Therefore,

$$\|g - f\|^2 = \|g\|^2 - 2\,\mathrm{Re}\,\langle g, \hat{f}\rangle + \|\hat{f}\|^2$$
$$= \|g\|^2 - 2\|\hat{g}\|^2 + \|f\|^2.$$

But $\|g\|^2 = \|\hat{g}\|^2$ and $\|f\|^2 = \|\hat{g}\|^2$. Therefore, $\|g - \hat{f}\|^2 = 0$, that is,

$$g(t) = \hat{f}(t). \tag{42}$$

In other words, every real (and, therefore, every complex) signal $g(t) \in L^2_{\mathbb{C}}(\mathbb{R})$ is the Fourier transform of some function of $L^2_{\mathbb{C}}(\mathbb{R})$. Hence, the mapping φ is *onto*. ■

EXERCISE **C3.5.** *Show that if a stable signal is base-band (that is, if its FT has compact support), then it also has a finite energy.*

We close this section by showing how the L^1 Fourier inversion theorem was limited in scope, since it does not take much for a stable signal not to have an integrable FT.

EXERCISE **C3.6.** *Show that if a stable signal is discontinuous at a point $t = a$, its FT is not integrable.*

C4

Fourier Series of Finite Power Periodic Signals

C4·1 Fourier Series in L^2_{loc}

Let us consider the Hilbert space $\ell^2_{\mathbb{C}}$ of complex sequences $a = \{a_n\}, n \in \mathbb{Z}$, such that $\sum_{n\in\mathbb{Z}} |a_n|^2 < \infty$, with the Hermitian product

$$\langle a, b\rangle_{\ell^2_{\mathbb{C}}} = \sum_{n\in\mathbb{Z}} a_n b_n^* \tag{43}$$

and the Hilbert space $L^2_{\mathbb{C}}([0, T], dt/T)$ of complex signals $x = \{x(t)\}, t \in \mathbb{R}$, such that $\int_0^T |x(t)|^2 \, dt < \infty$, with the Hermitian product

$$\langle x, y\rangle_{L^2_{\mathbb{C}}([0,T],\frac{dt}{T})} = \int_0^T x(t)y(t)^* \frac{dt}{T}. \tag{44}$$

THEOREM **C4.1.** *Formula*

$$\hat{s}_n = \frac{1}{T}\int_0^T s(t)e^{-2i\pi\frac{n}{T}t}\, dt \tag{45}$$

defines a linear isometry $s(\cdot) \to \{\hat{s}_n\}$ *from* $L^2_{\mathbb{C}}([0, T], dt/T)$ *onto* $\ell^2_{\mathbb{C}}$, *the inverse of which is given by*

$$s(t) = \sum_{n\in\mathbb{Z}} \hat{s}_n e^{2i\pi\frac{n}{T}t}, \tag{46}$$

where the series on the right-hand side converges in $L^2_{\mathbb{C}}([0, T], dt/T)$, *and the equality is almost everywhere. This isometry is summarized by the* Plancherel–

Parseval identity:

$$\sum_{n\in\mathbb{Z}} \hat{x}_n \hat{y}_n^* = \frac{1}{T}\int_0^T x(t)y(t)^*\, dt\,. \tag{47}$$

Proof: The result follows from general results on orthonormal bases of Hilbert spaces, since the sequence

$$\{e_n(\cdot)\} \stackrel{\text{def}}{=} \left\{\frac{1}{\sqrt{T}} e^{2i\pi\,\frac{n}{T}\cdot}\right\}, \quad n \in \mathbb{Z},$$

is a complete orthonormal sequence of $L^2_{\mathbb{C}}([0,T], dt/T)$ (Theorem C2.4). ■

The L^2 inversion theorem tells us that if $s(t)$ is a T-periodic complex signal with finite power, then

$$\sum_{-N}^{+N} \hat{s}_n e^{2i\pi(n/T)t} \xrightarrow[N\uparrow\infty]{L^2_{\mathbb{C}}([0,T])} s(t).$$

In general, for an arbitrary sequence of functions of $L^2_{\mathbb{C}}([0,T])$, convergence in $L^2_{\mathbb{C}}([0,T])$ does not imply convergence almost everywhere. However, for sequences of partial Fourier series, we have the surprising Carleson's theorem:

THEOREM **C4.2.** *The Fourier series of a T-periodic signal $s(t)$ with finite power converges almost everywhere to $s(t)$.*

This result shows that the situation for finite-power periodic signals is pleasant, in contrast with the situation prevailing locally stable periodic signals (remember Kolmogorov's result, Theorem A3.1). The proof of Carleson's result is omitted; it is rather technical. It also shows that the L^2 framework is very adapted to Fourier series, since everything works "as expected."

Discrete-Time Fourier Transform of Finite-Energy Signals

Let $\ell^1_{\mathbb{C}}$ be the space of sequences f_n, $n \in \mathbb{Z}$, such that $\sum_{n\in\mathbb{Z}} |f_n| < \infty$ (stable discrete-time signals).

THEOREM **C4.3.** *$\ell^1_{\mathbb{C}} \subset \ell^2_{\mathbb{C}}$, that is, a discrete-time stable signal has finite energy.*

Proof: Let $A = \{n : |x_n| \geq 1\}$. Since $\sum_{n\in\mathbb{Z}} |x_n| < \infty$, then necessarily $\text{card}(A) < \infty$. On the other hand, if $|x_n| \leq 1$, $|x_n|^2 \leq |x_n|$, whence

$$\sum_{n\in\mathbb{Z}} |x_n|^2 \leq \sum_{n\in A} |x_n|^2 + \sum_{n\in\overline{A}} |x_n| < \infty.$$

■

The situation for discrete-time signals is therefore in contrast with that of continuous-time signals, for which there exist stable signals with infinite energy that and finite-energy signals that are not stable.

Let $L^2_{\mathbb{C}}(2\pi)$ be the Hilbert space of functions $\tilde{f}:[-\pi,+\pi] \to \mathbb{C}$ such that $\int_{-\pi}^{+\pi} |\tilde{f}(\omega)|^2 \, d\omega < \infty$ provided with the Hermitian product

$$(\tilde{f}, \tilde{g})_{L^2_{\mathbb{C}}(2\pi)} = \frac{1}{2\pi} \int_{-\pi}^{+\pi} \tilde{f}(\omega)\tilde{g}(\omega)^* \, d\omega.$$

THEOREM **C4.4.** *There exists a linear isomorphism between $L^2_{\mathbb{C}}(2\pi)$ and $\ell^2_{\mathbb{C}}$ defined by*

$$f_n = \int_{-\pi}^{+\pi} \tilde{f}(\omega) e^{in\omega} \frac{d\omega}{2\pi}, \qquad \tilde{f}(\omega) = \sum_{n\in\mathbb{Z}} f_n e^{-in\omega}. \tag{48}$$

In particular, we have the Plancherel–Parseval identity

$$\sum_{n\in\mathbb{Z}} f_n g_n^* = \frac{1}{2\pi} \int_{-\pi}^{+\pi} \tilde{f}(\omega)\tilde{g}(\omega)^* \, d\omega. \tag{49}$$

Proof: This is a restatement of Theorem C4.3. ■

C4·2 Orthonormal Systems of Shifted Functions

We give a necessary and sufficient condition in the frequency domain for a system of shifted functions to be orthonormal.

THEOREM **C4.5.** *Let $g(t)$ be a function of $L^2_{\mathbb{C}}(\mathbb{R})$ and fix $0 < T < \infty$. A necessary and sufficient condition for the family of functions*

$$\{g(\cdot - nT)\}_{n\in\mathbb{Z}} \tag{50}$$

to form an orthonormal system of $L^2_{\mathbb{C}}(\mathbb{R})$ is

$$\sum_{n\in\mathbb{Z}} \left| \hat{g}\left(\nu + \frac{n}{T}\right) \right|^2 = T \quad \textit{almost everywhere.} \tag{51}$$

Proof: The Fourier transform $\hat{g}(\nu)$ of $g(t) \in L^2_{\mathbb{C}}(\mathbb{R})$ is in $L^2_{\mathbb{C}}(\mathbb{R})$ and, in particular, $|\hat{g}(\nu)|^2$ is integrable. By Theorem A2.3, $\sum_{n\in\mathbb{Z}} |\hat{g}(\nu + (n/T))|^2$ is $(1/T)$-periodic and locally integrable, and $T \int_{\mathbb{R}} g(t)\, g(t-nT)^* \, dt$ is its Fourier coefficient (this follows from the Plancherel–Parseval formula:

$$T \int_{\mathbb{R}} g(t) g(t-nT)^* \, dt = T \int_{\mathbb{R}} |\hat{g}(\nu)|^2 e^{-2i\pi\nu nT} \, d\nu$$

$$= T \int_{\mathbb{R}} \left\{ \sum_{k\in\mathbb{Z}} \left| \hat{g}\left(\nu + \frac{k}{T}\right) \right|^2 \right\} e^{-2i\pi\nu nT} \, d\nu).$$

The definition of orthonormality of system (50) is

$$\int_{\mathbb{R}} g(t)\, g(t-nT)^* \, dt = 1_{n=0}.$$

The proof then follows the argument in the proof of Theorem B2.6. ■

Riesz's Basis

The following notion will play an important role in multiresolution analysis.

DEFINITION **C4.1.** *A system of functions of* $L^2_{\mathbb{C}}(\mathbb{R})$

$$\{w(\cdot - nT)\}_{n\in\mathbb{Z}} \tag{52}$$

is said to form a Riesz basis *of some Hilbert subspace* V_0 *of* $L^2_{\mathbb{C}}(\mathbb{R})$ *if*

(a) it spans V_0, *and*

(b) for all sequence, $\{c_k\}_{k\in\mathbb{Z}}$ *of* $\ell^2_{\mathbb{C}}(\mathbb{Z})$,

$$A\sum_k |c_k|^2 \le \int_{\mathbb{R}} \left|\sum_{k\in\mathbb{Z}} c_k w(t-kT)\right|^2 dt \le B\sum_{k\in\mathbb{Z}} |c_k|^2, \tag{53}$$

where $0 < A \le B < \infty$ *are independent of the* c_k.

The function $\sum_{k\in\mathbb{Z}} c_k w(t-kT)$ has the Fourier transform $\sum_{k\in\mathbb{Z}} c_k e^{-2i\pi kT\nu}\widehat{w}(\nu)$, and, therefore, by the Plancherel–Parseval identity, the term between the bounds in (53) is equal to

$$\int_{\mathbb{R}} \left|\sum_{k\in\mathbb{Z}} c_k e^{-2i\pi kT\nu}\widehat{w}(\nu)\right|^2 d\nu$$

$$= \int_0^{1/T} \sum_{\ell\in\mathbb{Z}} \left|\sum_{k\in\mathbb{Z}} c_k e^{-2i\pi kT(\nu+\frac{\ell}{T})}\widehat{w}\left(\nu+\frac{\ell}{T}\right)\right|^2 d\nu$$

$$= \int_0^{1/T} \sum_{\ell\in\mathbb{Z}} \left|\sum_{k\in\mathbb{Z}} c_k e^{-2i\pi kT\nu}\widehat{w}\left(\nu+\frac{\ell}{T}\right)\right|^2 d\nu$$

$$= \int_0^{1/T} \left|\sum_{k\in\mathbb{Z}} c_k e^{-2i\pi kT\nu}\right|^2 \sum_{\ell\in\mathbb{Z}} \left|\widehat{w}\left(\nu+\frac{\ell}{T}\right)\right|^2 d\nu.$$

Also,

$$T\int_0^{1/T} \left|\sum_{k\in\mathbb{Z}} c_k e^{-2i\pi kT\nu}\right|^2 d\nu = \sum_{k\in\mathbb{Z}} |c_k|^2.$$

Now, any function $c(\nu) \in Ł^2_{\mathbb{C}}([0,1/T])$ has the form $\sum_{k\in\mathbb{Z}} c_k e^{-2i\pi kT\nu}$, where $\sum_{k\in\mathbb{Z}} |c_k|^2 < \infty$, and (53)is therefore equivalent to

$$AT\int_0^{1/T} |c(\nu)|^2 d\nu \le \int_0^{1/T} |c(\nu)|^2 \left\{\sum_{\ell\in\mathbb{Z}} \left|\widehat{w}\left(\nu+\frac{\ell}{T}\right)\right|^2\right\} d\nu$$

$$\le BT\int_0^{1/T} |c(\nu)|^2 d\nu$$

for any $c(\nu) \in L^2_{\mathbb{C}}([0, 1/T])$. It then follows that

$$AT \le \sum_{n\in\mathbb{Z}} \left|\widehat{w}\left(\nu + \frac{\ell}{T}\right)\right|^2 \le BT \quad \text{a.e.} \tag{54}$$

Theorem C4.6. *Let $\{w(\cdot - nT)\}_{n\in\mathbb{Z}}$ be a Riesz basis of some Hilbert subspace $V_0 \subset L^2_{\mathbb{C}}(\mathbb{R})$. Define the function $g \in L^2_{\mathbb{C}}(\mathbb{R})$ by its Fourier transform*

$$\hat{g}(\nu) = \sqrt{T}\,\frac{\widehat{w}(\nu)}{\left(\sum_{n\in\mathbb{Z}} \left|\widehat{w}\left(\nu + \frac{\ell}{T}\right)\right|^2\right)^{1/2}}\,. \tag{55}$$

Then $\{g(\cdot - nT)\}_{n\in\mathbb{Z}}$ is a Hilbert basis of V_0.

Proof: In view of (54), the function g is well defined and in $L^2_{\mathbb{C}}(\mathbb{R})$. Since (51) is obviously satisfied, it follows that the system $\{g(\cdot - nT)\}_{n\in\mathbb{Z}}$ is orthonormal.

We must now show that the Hilbert space $\widetilde{V}_0$ spanned by $\{g(\cdot - nT)\}_{n\in\mathbb{Z}}$ is in fact identical to V_0. It suffices to show that the generators of V_0 belong to $\widetilde{V}_0$, and vice versa. Define

$$\alpha(\nu) = \sqrt{T}\left(\sum_{n\in\mathbb{Z}} \left|\widehat{w}\left(\nu + \frac{\ell}{T}\right)\right|^2\right)^{-1/2}.$$

In view of condition (54), $\alpha(\nu)$ is $(1/T)$-periodic and of finite power, and it therefore admits a Fourier representation

$$\alpha(\nu) = \sum_{n\in\mathbb{Z}} \alpha_n e^{-2i\pi\nu nT},$$

for some sequence $\{\alpha_n\} \in \ell^2_{\mathbb{C}}(\mathbb{Z})$. Since

$$\widehat{w}(\nu) = \alpha(\nu)\hat{g}(\nu),$$

it follows that

$$w(t) = \sum_{n\in\mathbb{Z}} \alpha_n g(t - nT).$$

Therefore, the generators of V_0 are in $\widetilde{V}_0$. The converse is true by the same argument since

$$\hat{g}(\nu) = \alpha(\nu)^{-1}\widehat{w}(\nu),$$

where $\alpha(\nu)^{-1}$ is also, in view of condition (54), a $(1/T)$-periodic of finite power. ■

Exercise C4.1. *Let $h(t)$ be the function of FT $(1/\sqrt{2B})1_{[-B,+B]}(\nu)$ for some $B > 0$. Show that there exists no orthonormal basis of $L^2_{\mathbb{C}}(\mathbb{R}; B)$ of the form $\left\{\frac{1}{\sqrt{2B}}\, g\left(t - \frac{n}{2B}\right)\right\}_{n\in\mathbb{Z}}$, where*

$$g(t) = \sum_{k\in\mathbb{Z}} c_k h\left(t - \frac{k}{2B}\right),$$

where $\{c_n\}_{n\in\mathbb{Z}}$ *is in* $\ell^2_{\mathbb{C}}(\mathbb{Z})$, *unless only one of the* c_n *is nonzero. Show that if the Fourier sum* $\tilde{c}(\omega)$ *of the sequence* $\{c_n\}_{n\in\mathbb{Z}}$ *is such that* $A < |\tilde{c}(\omega)|^2 < B$ *for some* $0 < A \leq B < \infty$ *and for all* ω, *then* $\left\{\frac{1}{\sqrt{2B}}\, g\left(t - \frac{n}{2B}\right)\right\}_{n\in\mathbb{Z}}$ *is a Riesz basis of* $L^2_{\mathbb{C}}(\mathbb{R}; B)$.

References

[C1] Daubechies, I. (1992). *Ten Lectures on Wavelets, CBSM–NSF Regional Conf. Series in Applied Mathematics*, SIAM: Philadelphia, PA.

[C2] Gasquet, C. and Witomski, P. (1991). *Analyse de Fourier et Applications*, Masson: Paris.

[C3] Halmos, P.R. (1951). *Introduction to Hilbert space*, Chelsea, New York.

[C4] Rudin, W. (1966). *Real and Complex Analysis*, McGraw-Hill, New York.

[C5] Young, N.Y. (1988). *An Introduction to Hilbert Spaces*, Cambridge University Press.

Part D

Wavelet Analysis

Introduction

Although Fourier theory had reached in the L^2 framework a formal mathematical beauty, it was not entirely satisfactory for important applications in signal processing. Indeed, in many situations, the information contained in a signal is localized in time and/or in frequency. The typical example is a piece of music, which is perceived as a succession of notes well localized in both time and frequency. The usual Fourier transform is not adapted to the analysis of music because for a given frequency (a note) it is related to the total energy of all occurrences of this note in the entire piece.

This led Dennis Gabor[1] to propose a windowed Fourier transform, whose idea is very natural. If $f(t)$ is the signal to be analyzed, the local information at time $t = b$ is contained in the time-localized signal

$$f(t)w^*(t-b),$$

where $w(t)$ is the *time window function*, a function negligible outside a relatively small interval around zero.

Given a window $w(t)$, the local information at time b is obtained by computing the Fourier transform of the last display:

$$W_f(\nu, b) := \int_{\mathbb{R}} f(t)w^*(t-b)e^{-2i\pi\nu t}\,dt = \langle f, w_{\nu,b}\rangle,$$

where

$$w_{\nu,b}(t) = w(t-b)e^{2i\pi\nu t}.$$

[1]Theory of Communications, *J. Inst. Elec. Engrg.*, Vol. 93, pp. 429–457, 1946.

We see that the time information is collected in a time interval around time b of width of the order of that of the time window. Now, from the Plancherel–Parseval identity,

$$W_f(\nu, b) = e^{-2i\pi\nu b} \int_{\mathbb{R}} \hat{f}(\mu)\widehat{w}(\mu - \nu)^* e^{2i\pi\mu b}\, \mathrm{d}\mu.$$

Therefore, we see that the frequency information is collected in a frequency interval around ν of the width of the time window's FT.

These observations point to a fundamental limitation of the windowed Fourier transform, in relation with the uncertainty principle, which states that time resolution is possible only at the expense of frequency resolution, and vice versa. Indeed, the wider the function, the narrower its Fourier transform, and vice versa. Of course, this is an imprecise statement, but it is already substantiated by the Doppler theorem

$$|a|^{\frac{1}{2}} f(at) \overset{\mathrm{FT}}{\to} \frac{1}{|a|^{\frac{1}{2}}} \hat{f}\left(\frac{\nu}{|a|}\right).$$

(Note that as a varies, the energy of $|a|^{\frac{1}{2}} f(at)$ remains the same.) The so-called Heisenberg uncertainty principle makes the above limitation more explicit and states that

$$\sigma_w \sigma_{\widehat{w}} \geq \frac{1}{4\pi},$$

where σ_f is the mean-square width of a function $f \in L^2_{\mathbb{C}}(\mathbb{R})$ (see the definition in the first lines of Section D1·1). The conclusion is that as long as we resort to windowed Fourier transforms, time resolution and frequency resolution are antagonistic: If one is increased, the other is necessarily decreased, keeping the area $\sigma_w \sigma_{\widehat{w}}$ of the time–frequency box above the lower bound of $1/4\pi$. However, the real inconvenience of the windowed FT is the fixed shape of the time–frequency box. In many occasions, it is interesting to have a time–frequency box that adapts itself to the time–frequency point analyzed. For instance, a discontinuity (abrupt change) of a signal takes place in a short time and involves high frequencies. Therefore, at high frequencies, the time dimension of the time–frequency box should be small. Also, since it takes time to determine the frequency of a low-frequency sinusoid, at low frequencies the time dimension of the time–frequency box should be large. Motivated by these imperatives, Jean Morlet[2] proposed the wavelet transform in order to take into account the need for an adaptive time–frequency box.

[2] Sampling theory and wave propagation, in NATO ASI series, Vol. 1, *Survey in Acoustic Signal/ Image Processing and Recognition*, C.H. Chen, ed., Springer-Verlag, Berlin, 233–261, 1983.

In the wavelet transform, the role the family of functions $w_{\nu,b}(t)$ plays in the windowed Fourier transform is played by a family

$$\psi_{a,b}(t) = |a|^{-1/2}\psi\left(\frac{t-b}{a}\right), \quad a, b \in \mathbb{R},\ a \neq 0,$$

where $\psi(t)$ is called the *mother wavelet.* The wavelet transform (WT) of the function $f \in L^2_{\mathbb{C}}(\mathbb{R})$ is the function

$$C_f(a, b) = \langle f, \psi_{a,b}\rangle = \int_{\mathbb{R}} f(t)\psi^*_{a,b}(t)\,\mathrm{d}t, \quad a, b \in \mathbb{R},\ a \neq 0.$$

By the Plancherel–Parseval identity,

$$C_f(a, b) = \langle \hat{f}, \widehat{\psi}_{a,b}\rangle = \int_{\mathbb{R}} \hat{f}(\nu)\widehat{\psi}^*_{a,b}(\nu)\,\mathrm{d}\nu,$$

where

$$\widehat{\psi}_{a,b}(\nu) = |a|^{1/2}e^{-2i\pi\nu b}\widehat{\psi}(a\nu).$$

From the above expression, it appears that the wavelet transform $C_f(a, b)$ analyzes the function $f(t)$ in the time–frequency box

$$[b + am - a\sigma, b + am + a\sigma] \times \left[\frac{\widehat{m}}{a} - \frac{\widehat{\sigma}}{a}, \frac{\widehat{m}}{a} + \frac{\widehat{\sigma}}{a}\right]$$

(where m is the center of ψ, σ is its width; $\widehat{m}$ and $\widehat{\sigma}$ are the corresponding objects relatively to $\widehat{\psi}$; see the details in the main text). Assuming $\widehat{m} > 0$, we see that the frequency window is centered at $\nu = \widehat{m}/a$ and has width $2\widehat{\sigma}/a$; therefore, the ratio center/width

$$Q = \frac{\widehat{m}}{2\widehat{\sigma}}$$

is independent of the frequency variable a. The area of the time–frequency box is constant, but its shape varies with the frequency $\nu = \widehat{m}/a$ analyzed. For high frequencies it has a large time dimension, and for small frequencies it has a small time dimension, which is the desired effect.

The interest of the wavelet transform for signal analysts is that they can "read" it to extract information about the time–frequency structure that is otherwise blurred in the brute signal by concurrent phenomena and subsidiary effects. For instance, they can detect the appearance time of a phenomenon linked to a particular frequency (e.g., the time at which a particular atom starts to be excited).

A formula, called the *identity resolution*, allows us to reconstruct, under mild conditions that we shall make precise in due time, the function from its wavelet transform:

$$f(t) = \frac{1}{K}\int_{\mathbb{R}}\int_{\mathbb{R}} C_f(a, b)\psi_{a,b}(t)\,\frac{\mathrm{d}a\,\mathrm{d}b}{a^2},$$

where K is some constant depending on ψ. In this sense, wavelet analysis is *continuous*, in that the original function of L^2 is reconstructed as a continuous linear combination of the continuous wavelet basis $\psi_{a,b}(t) = |a|^{-1/2}\psi\left(\frac{t-b}{a}\right)$, $a, b \in \mathbb{R}$, $a \neq 0$. One would rather store the original function not as a function of two arguments, but as the doubly indexed sequence of coefficients of a decomposition along an orthonormal base of L^2, $\{\psi_{j,n}\}_{j,n\in\mathbb{Z}}$, called the *wavelet basis*,

$$f = \sum_{j\in\mathbb{Z}} \sum_{k\in\mathbb{Z}} \langle f, \psi_{j,k}\rangle \psi_{j,k},$$

where

$$\psi_{j,n}(t) = 2^{\frac{j}{2}}\psi(2^j t - n),$$

and where $\psi(t)$ is the *mother wavelet*. The multiresolution analysis of Stéphane Mallat[3] is one particular way of obtaining such orthonormal bases and is the main topic of Part D. Similarly to the continuous wavelet transform, the coefficients of the multiresolution decomposition can be used to analyze a signal. But multiresolution analysis is also a tool for data compression. Indeed, with a good design of the mother wavelet, many wavelet coefficients are small and can be neglected. The coefficients that are not neglected are quantized more or less coarsely, depending, for instance, on the frequency index. This is of course reminiscent of subband coding, and this resemblance is not at all a coincidence. Mallat's algorithm of analysis–synthesis is of the same form as the subband analysis–synthesis algorithm. Multiresolution analysis can be considered as a systematic way of doing subband analysis. One of its advantages is to place the latter in a framework where the mathematical issues ensuring the efficiency of the algorithms are more easily dealt with.

Perhaps one of the most striking advantages of multiresolution analysis over classical Fourier analysis is in the way it handles discontinuities. Consider, for instance, the signal on top of Figs. D0.1a and D0.1b,[4] which has a spike. In both figures the middle signal is a Fourier series approximation of the top signal, whereas the bottom signal is a wavelet approximation of the top signal. In Fig. D0.1a, only the first 60 coefficients of the expansions (Fourier or wavelet) are used to produce the approximated signals, and it appears that there is not a dramatic difference between the two approximations. This is not the case, however, when, as in Fig. D0.1b, one uses the 60 *largest* coefficients. The advantage of the wavelet approximation is then obvious.

[3]Multiresolution approximation and wavelets, *Trans. Amer. Math. Soc.*, **315**, 1989, 69–88.

[4]Reproduced with the kind permission of Martin Vetterli.

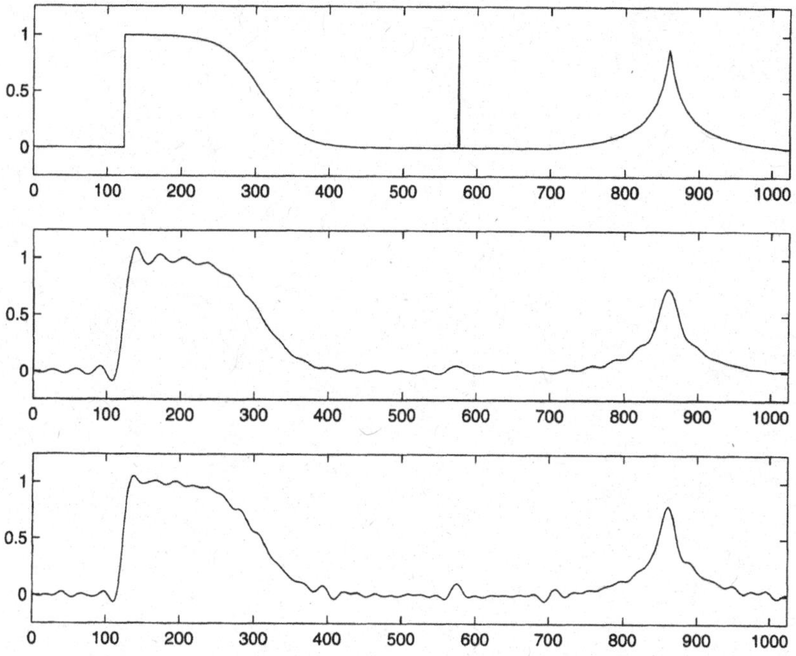

(a)

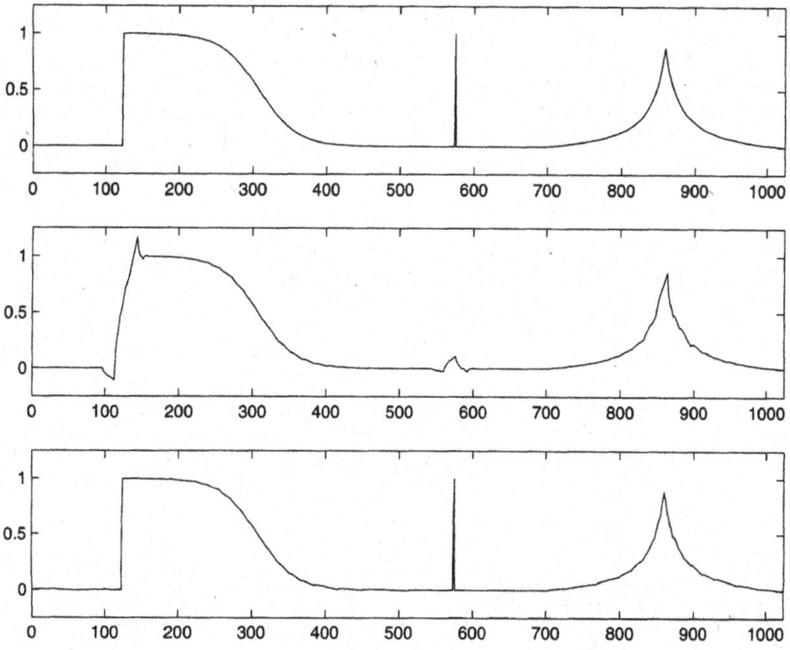

(b)

Figure D0.1. Wavelet vs. Fourier

D1

The Windowed Fourier Transform

D1·1 The Uncertainty Principle

Fourier analysis, well as wavelet analysis have an intrinsic limitation, which is contained in the *uncertainty principle.* In order to state this result, we need a definition of the "width" of a function. Here is the one that suits our purpose.

Root Mean-Square Widths

Let $w : \mathbb{R} \mapsto \mathbb{C}$ be a nontrivial function in L^2. Define the centers of w and $\widehat{w}$, respectively, by

$$m_w = \frac{1}{E_w} \int_{\mathbb{R}} t\,|w(t)|^2\,dt\,,$$

$$m_{\widehat{w}} = \frac{1}{E_w} \int_{\mathbb{R}} \nu\,|w(\nu)|^2\,d\nu,$$

where E_w is the energy of $w(t)$:

$$E_w = \int_{\mathbb{R}} |w(t)|^2\,dt = \int_{\mathbb{R}} |\widehat{w}(\nu)|^2\,d\nu.$$

Then define

$$\sigma_w = \left(\frac{1}{E_w} \int_{\mathbb{R}} |t - m_w|^2\,|w(t)|^2\,dt\right)^{\frac{1}{2}}$$

and

$$\sigma_{\widehat{w}} = \left(\frac{1}{E_w} \int_{\mathbb{R}} |\nu - m_{\widehat{w}}|^2\,|\widehat{w}(\nu)|^2\,d\nu\right)^{\frac{1}{2}}.$$

The numbers σ_w and $\sigma_{\widehat{w}}$ are the root mean-square (RMS) widths of w and $\widehat{w}$, respectively. Note that m_w and $m_{\widehat{w}}$ are not always defined. When they are well defined, $\sigma_{\widehat{w}}$ and $\sigma_{\widehat{w}}$ are always defined but may be infinite. Therefore, we shall always assume that

$$\int_{\mathbb{R}} |t|\,|w(t)|^2\,dt < \infty$$

and

$$\int_{\mathbb{R}} |\nu|\,|w(\nu)|^2\,d\nu < \infty,$$

to guarantee at least the existence of the centers of $w(t)$ and of its FT.

EXERCISE **D1.1.** *Check if the centers m_w and $m_{\widehat{w}}$ are well defined, and then compute σ_w and $\sigma_{\widehat{w}}$ and the product $\sigma_w\sigma_{\widehat{w}}$ in the following cases:*

$$\begin{aligned} w(t) &= 1_{[-T,+T]} \\ w(t) &= e^{-a|t|}, \quad a > 0, \\ w(t) &= e^{-at^2}, \quad a > 0. \end{aligned}$$

EXERCISE **D1.2.** *Suppose that the center m_w of the function $w \in L^2$ is well defined. Show that the quantity*

$$\int_{\mathbb{R}} |t - t_0|^2\,|w(t)|^2\,dt$$

is minimized by $t_0 = m_w$.

Heisenberg's Inequality

THEOREM **D1.1.** *Under the conditions stated above, we have* Heisenberg's inequality

$$\sigma_w\sigma_{\widehat{w}} \geq \frac{1}{4\pi}\,. \tag{1}$$

Proof: We assume that the window and its FT are centered at 0, without loss of generality (see Exercise D1.3). Denoting the L^2-norm of a function f by $\|f\|$, we have to show that

$$\|tw\| \times \|\nu\widehat{w}\| \geq \frac{1}{4\pi}\,\|w\|^2\,. \tag{2}$$

We first assume that w is a C^∞-function with bounded support. In particular,

$$\widehat{w'}(\nu) = (2i\pi\nu)\,\widehat{w}(\nu),$$

and therefore,

$$\|\nu\widehat{w}\|^2 = \frac{1}{4\pi^2}\,\|\widehat{w'}\|^2 = \frac{1}{4\pi^2}\,\|w'\|^2.$$

Thus, it remains to show that

$$\|tw\| \times \|w'\| \geq \frac{1}{2}\|w\|^2 . \tag{3}$$

By Schwarz's inequality,

$$\|tw\| \times \|w'\| \geq |\langle tw, w'\rangle| \geq |\mathrm{Re}\{\langle tw, w'\rangle\}|.$$

Now,

$$2\mathrm{Re}\{\langle tw, w'\rangle\} = \langle tw, w'\rangle + \langle w', tw\rangle = \int_{\mathbb{R}} t(ww'^* + w'w^*)\,dt$$

$$= t|w(t)|^2|_{-\infty}^{+\infty} - \int_{\mathbb{R}} |w(t)|^2\,dt = 0 - \|w\|^2.$$

This gives (3) in the case where $w \in C^\infty$. We now show that (3) is true in the general case. To see this, we first observe that it suffices to prove (3) in the case where w belongs to the Hilbert space

$$H = \{w \in L^2_{\mathbb{C}}(\mathbb{R});\ tw(t) \in L^2_{\mathbb{C}}(\mathbb{R}),\ \nu\widehat{w}(\nu) \in L^2_{\mathbb{C}}(\mathbb{R})\},$$

with the norm $\|w\|_H = \left(\|w\|^2 + \|tw\|^2 + \|\nu\widehat{w}\|^2\right)^{\frac{1}{2}}$ (if $w(t)$ is not in this space, Heisenberg's inequality is trivially satisfied). Then we use the fact that the subset of H consisting of the C^∞-functions with compact support is dense in H (see Exercise C3.4). The result then follows from the continuity of the Hermitian product. ■

Equality in (1) takes place if and only if $w(t)$ is proportional to a Gaussian signal e^{-ct^2}, where $c > 0$. We do the proof only in the case where it is further assumed that $w \in S$ and is real. Observe that all the steps in the first part of the proof remain valid since for such a function, $t|w(t)|^2|_{-\infty}^{+\infty} = 0$. Equality is attained in (1) if and only if the functions $tw(t)$ and $w'(t)$ are proportional, say,

$$w'(t) = -ctw(t),$$

and this gives

$$w(t) = Ae^{-ct^2},$$

where $c > 0$ necessarily, because $w(t) \in L^2$.

EXERCISE **D1.3.** *Show that, for arbitrary $t_0 \in \mathbb{R}$, $\nu_0 \in \mathbb{R}$,*

$$\|(t - t_0)w\| \times \|(\nu - \nu_0)\widehat{w}\| \geq \frac{1}{4\pi}\|w\|^2 . \tag{4}$$

(Hint: Consider the function $g(t) = e^{-2i\pi\nu_0 t}w(t + t_0)$.)

The above result tells us that in Heisenberg's inequality (1), the numbers σ_w and $\sigma_{\widehat{w}}$ can be taken to be, respectively, the root mean-square width of w around any time t_0 and the root mean-square width of $\widehat{w}$ around any frequency ν_0. In particular, t_0 can be the center of $w(t)$, and ν_0 can be the center of $\widehat{w}(\nu)$. This version is the most stringent, since the RMS widths around the centers are the smallest RMS widths (see Exercise D1.2).

D1.2 The WFT and Gabor's Inversion Formula

Windows

Let $f(t)$ be the signal to be analyzed. The local information at (around) time $t = b$ is contained in the time-localized signal

$$f(t)w^*(t-b), \tag{5}$$

where $w(t)$ is the time *window function*, the support of which is included in a relatively small interval about zero. Typical examples are the *rectangular window*

$$w(t) = 1_{[-\alpha,+\alpha]}(t)$$

and the *Gaussian window*

$$w(t) = e^{-\alpha t^2},$$

where $\alpha > 0$.

Given a window $w(t)$, the local information at time b is obtained by computing the Fourier transform of (5):

$$W_f(\nu, b) := \int_{\mathbb{R}} f(t)w^*(t-b)e^{-2i\pi\nu t}\,dt. \tag{6}$$

The choice of w^* instead of w in (5) is purely for notational comfort. For example,

$$W_f(\nu, b) = \langle f, w_{\nu,b}\rangle, \tag{7}$$

the Hermitian product in $L^2_{\mathbb{C}}(\mathbb{R})$ of $f(t)$ with

$$w_{\nu,b}(t) = w(t-b)e^{2i\pi\nu t}, \tag{8}$$

where we assume, of course, that f and w are complex-valued functions that have finite energy.

DEFINITION **D1.1.** *Let $w, f \in L^2$. One calls the function $W_f : \mathbb{R} \times \mathbb{R} \to \mathbb{C}$ defined by* (6) *the* windowed Fourier transform *(or WFT) of f associated with the window w.*

When the rectangular window is chosen, W_f is called the *short-time Fourier transform* of f. If the window is Gaussian, the function W_f is called the *Gabor transform* of f.

Inversion Formula

THEOREM **D1.2.** *Under the following assumptions,*

(a) $w \in L^1_{\mathbb{C}}(\mathbb{R}) \cap L^2_{\mathbb{C}}(\mathbb{R})$,

(b) $\int_{\mathbb{R}} |w(t)|^2\,dt = 1,$

(c) $|\widehat{w}|$ is an even function,

we have, for all $f \in L^2_{\mathbb{C}}(\mathbb{R})$, the energy conservation formula

$$\int_{\mathbb{R}}\int_{\mathbb{R}} |W_f(\nu, b)|^2 \, d\nu \, db = \int_{\mathbb{R}} |f(t)|^2 \, dt, \tag{9}$$

and the reconstruction formula, *(*inversion formula*),*

$$\lim_{A\uparrow\infty} \int_{\mathbb{R}} \left| f(t) - \int_{|\nu|\le A} \int_{\mathbb{R}} W_f(\nu, b) w_{\nu,b}(t) \, d\nu \, db \right|^2 dt = 0. \tag{10}$$

Note that assumption (c) is automatically satisfied if $w(t)$ is a real function. Assumption (b) is just a convention: The window can always be normalized to have an energy equal to 1, and the wavelet transform is then just multiplied by a constant.

Proof: The proof is technical and can be skipped in a first reading. We define

$$f_A(t) = \int_{|\nu|\le A} \int_{\mathbb{R}} W_f(\nu, b) w_{\nu,b}(t) \, d\nu \, db.$$

Using the Plancherel–Parseval identity, (7) becomes

$$W_f(\nu, b) = \langle \hat{f}, \widehat{w}_{\nu,b} \rangle,$$

where

$$\widehat{w}_{\nu,b}(\mu) = e^{-2i\pi(\mu-\nu)b} \widehat{w}(\mu - \nu) \tag{11}$$

is the Fourier transform of (8). Therefore,

$$W_f(\nu, b) = e^{-2i\pi\nu b} \int_{\mathbb{R}} \hat{f}(\mu) \widehat{w}(\mu - \nu)^* e^{2i\pi\mu b} \, d\mu. \tag{12}$$

The function $\mu \to \hat{f}(\mu)\widehat{w}(\mu - \nu)^*$ is in $L^1 \cap L^2$. (It is in L^1 as the product of two L^2-functions; it is in L^2 because $\hat{f} \in L^2$ and $\widehat{w}$ is bounded, being the Fourier transform of an L^1-function.) By the Plancherel–Parseval identity,

$$\begin{aligned}\int_{\mathbb{R}} |W_f(\nu, b)|^2 \, db &= \int_{\mathbb{R}} \left| \int_{\mathbb{R}} \hat{f}(\mu) \widehat{w}(\mu - \nu)^* e^{2i\pi\mu b} \, d\mu \right|^2 db \\ &= \int_{\mathbb{R}} |\hat{f}(\mu) \widehat{w}(\mu - \nu)^*|^2 \, d\mu,\end{aligned}$$

and, therefore,

$$\begin{aligned}\int_{\mathbb{R}}\int_{\mathbb{R}} |W_f(\nu, b)|^2 \, db \, d\nu &= \int_{\mathbb{R}}\int_{\mathbb{R}} |\hat{f}(\mu)|^2 |\widehat{w}(\mu - \nu)|^2 \, d\mu \, d\nu \\ &= \int_{\mathbb{R}} \left\{ |\hat{f}(\mu)|^2 \int_{\mathbb{R}} |\widehat{w}(\mu - \nu)|^2 \, d\nu \right\} d\mu \\ &= \int_{\mathbb{R}} \left\{ |\hat{f}(\mu)|^2 \int_{\mathbb{R}} |\widehat{w}(\nu)|^2 \, d\nu \right\} d\mu.\end{aligned}$$

Equality (9) follows because

$$\int_{\mathbb{R}} |\widehat{w}(\nu)|^2 \, d\nu = \int_{\mathbb{R}} |w(t)|^2 \, dt = 1,$$

$$\int_{\mathbb{R}} |\hat{f}(\mu)|^2 \, d\mu = \int_{\mathbb{R}} |f(t)|^2 \, dt \quad \text{(Plancherel–Parseval identity).}$$

We show that the function f_A is well defined, that is, $(\nu, b) \to W_f(\nu, b) w_{\nu,b}(t)$ is integrable over $[-A, +A] \times \mathbb{R}$. In view of (12),

$$\begin{aligned} I_A(t) &:= \int_{-A}^{+A} \int_{\mathbb{R}} |W_f(\nu, b)| \, |w_{\nu,b}(t)| \, d\nu \, db \\ &= \int_{-A}^{+A} \left(\int_{\mathbb{R}} \left| \mathcal{F}^{-1}\{\hat{f}(\cdot)\widehat{w}(\cdot - \nu)^*\}(b) \right| |w(t-b)| \, db \right) d\nu. \end{aligned}$$

By Schwarz's inequality and the Plancherel–Parseval identity, and using assumption (b),

$$\begin{aligned} &\int_{\mathbb{R}} \left| \mathcal{F}^{-1}\{\hat{f}(\cdot)\widehat{w}(\cdot - \nu)^*\}(b) \right| |w(t-b)| \, db \\ &\leq \left(\int_{\mathbb{R}} \left| \mathcal{F}^{-1}\{\hat{f}(\cdot)\widehat{w}(\cdot - \nu)^*\} \right|^2 db \right)^{1/2} \left(\int_{\mathbb{R}} |w(t-b)|^2 \, db \right)^{1/2} \\ &= \left(\int_{\mathbb{R}} \left| \mathcal{F}^{-1}\{\hat{f}(\cdot)\widehat{w}(\cdot - \nu)^*\} \right|^2 db \right)^{1/2} \\ &= \int_{\mathbb{R}} |\hat{f}(\mu)\widehat{w}(\mu - \nu)^*|^2 \, d\mu \\ &= \left(|\hat{f}|^2 * |\widehat{w}|^2 \right)(\nu) := h(\nu). \end{aligned}$$

This function $h(\nu)$ is in L^1, being the convolution product of two L^1-functions. In particular,

$$I_A(t) \leq \int_{-A}^{+A} |h(\nu)| \, d\nu < \infty,$$

and f_A is therefore well defined. Using a previous calculation, we have

$$f_A(t) = \int_{-A}^{+A} g(\nu) \, d\nu,$$

where

$$g(\nu) := \int_{\mathbb{R}} \mathcal{F}^{-1}\{\hat{f}(\cdot)\widehat{w}(\cdot - \nu)^*\}(b) w(t-b) e^{2i\pi\nu(t-b)} \, db.$$

By the Plancherel–Parseval identity,

$$g(\nu) = \int_{\mathbb{R}} \hat{f}(\mu)\widehat{w}(\mu-\nu)^*\widehat{w}(\mu-\nu)e^{2i\pi\mu t}\,d\mu$$

$$= \int_{\mathbb{R}} \hat{f}(\mu)|\widehat{w}(\mu-\nu)|^2 e^{2i\pi\mu t}\,d\mu.$$

Therefore,

$$f_A(t) = \int_{-A}^{+A}\left(\int_{\mathbb{R}} \hat{f}(\mu)|\widehat{w}(\mu-\nu)|^2 e^{2i\pi\mu t}\,d\mu\right)d\nu.$$

In order to change the order of integration in the above integral, we first verify that $(\nu,\mu)\to|\hat{f}(\mu)|\,|\widehat{w}(\mu-\nu)|^2$ is integrable over $[-A,+A]\times\mathbb{R}$. But $|\hat{f}|\in L^2$ and $|\widehat{w}|^2\in L^1$; therefore, $|\hat{f}| * |\widehat{w}|^2\in L^2$, and the integral of an L^2-function over a finite interval is finite. But this integral is just

$$\int_{-A}^{+A}\left(\int_{\mathbb{R}} |\hat{f}(\mu)|\,|\widehat{w}(\nu-\mu)|^2\,d\mu\right)d\nu = \int_{-A}^{+A}\left(\int_{\mathbb{R}} |\hat{f}(\mu)|\,|\widehat{w}(\mu-\nu)|^2\,d\mu\right)d\nu$$

since $|\widehat{w}|$ is even. We can now apply Fubini's theorem to obtain

$$f_A(t) = \int_{\mathbb{R}}\left(\int_{-A}^{+A} \hat{f}(\mu)|\widehat{w}(\mu-\nu)|^2 e^{2i\pi\mu t}\,d\nu\right)d\mu$$

$$= \int_{\mathbb{R}} \hat{f}(\mu)e^{2i\pi\mu t}\left(\int_{-A}^{+A} |\widehat{w}(\mu-\nu)|^2\,d\nu\right)d\mu$$

$$= \int_{\mathbb{R}} \hat{f}(\mu)\varphi_A(\mu)e^{2i\pi\mu t}\,d\mu,$$

where $0\le\varphi_A(\mu)\le 1$, in view of assumption (b). In particular, $\hat{f}\varphi_A\in L^2$, and

$$f_A = \mathcal{F}^{-1}(\hat{f}\varphi_A).$$

We now show that $\lim_{A\uparrow\infty} f_A = f$ in L^2. For this, we write, using the Plancherel–Parseval identity,

$$\|f-f_A\|^2_{L^2} = \|\mathcal{F}^{-1}\hat{f}-\mathcal{F}^{-1}\hat{f}\varphi_A\|^2_{L^2}$$

$$= \|\mathcal{F}^{-1}\{\hat{f}(1-\varphi_A)\}\|^2_{L^2}$$

$$= \|\hat{f}(1-\varphi_A)\|^2_{L^2}$$

$$= \int_{-\infty}^{+\infty} |1-\varphi_A(\mu)|^2|\hat{f}(\mu)|^2\,d\mu.$$

We have

$$1-\varphi_A(\mu) = \int_{|\nu|\ge A} |\widehat{w}(\mu-\nu)|^2\,d\nu$$

$$= \int_{-\infty}^{\mu-A} |\widehat{w}(y)|^2\,dy + \int_{\mu+A}^{+\infty} |\widehat{w}(y)|^2\,dy,$$

and, therefore, if $|\mu| \leq A/2$,

$$\begin{aligned} 0 &\leq 1 - \varphi_A(\mu) \\ &\leq \int_{-\infty}^{-A/2} |\widehat{w}(y)|^2\, dy + \int_{A/2}^{+\infty} |\widehat{w}(y)|^2\, dy \\ &= \gamma(A) \to 0 \quad (A \to \infty). \end{aligned}$$

Therefore,

$$\int_{-A/2}^{+A/2} |1 - \varphi_A(\mu)|^2 |\hat{f}(\mu)|^2\, d\mu \leq \gamma(A) \|f\|_{L^2}^2 \to 0 \quad (A \to \infty).$$

Also,

$$\begin{aligned} \int_{|\mu| \geq A/2} |1 - \varphi_A(\mu)|^2 |\hat{f}(\mu)|^2\, d\mu &\leq \int_{|\mu| \geq A/2} |\hat{f}(\mu)|^2\, d\mu \\ &\to 0 \quad (A \to \infty) \end{aligned}$$

since $\hat{f} \in L^2$. Therefore, finally,

$$\lim_{A \uparrow \infty} \|f - f_A\|_{L^2}^2 = 0. \qquad \blacksquare$$

From (7) and the Plancherel–Parseval identity, we obtain the two expressions for the windowed Fourier transform:

$$W_f(\nu, b) = \langle f, w_{\nu,b} \rangle = \langle \hat{f}, \widehat{w}_{\nu,b} \rangle, \tag{13}$$

where $\widehat{w}_{\nu,b}$ is defined by (11). We assume (without diminishing the generality of the discussion to follow) that w and $\widehat{w}$ are functions centered at zero, that is,

$$\int_{\mathbb{R}} t|w(t)|^2\, dt = 0 \quad \text{and} \quad \int_{\mathbb{R}} \nu |\widehat{w}(\nu)|^2\, d\nu = 0.$$

The RMS width σ_w is an indicator of the localization of $w_{\nu,b}$ about b, whereas $\sigma_{\widehat{w}}$ is an indicator of the localization of $\widehat{w}_{\nu,b}$ about ν. The rectangle $[b - \sigma_w, b + \sigma_w] \times [\nu - \sigma_{\widehat{w}}, \nu + \sigma_{\widehat{w}}]$ is the local time–frequency box about (b, ν) analyzed by the windowed Fourier transform at (b, ν).

It is of interest to have a sharp resolution, that is, to make the area $4\sigma_w \sigma_{\widehat{w}}$ of the time–frequency box as small as possible. However, windows have the basic limitation contained in the uncertainty principle, which says that

$$\sigma_w \sigma_{\widehat{w}} \geq \frac{1}{4\pi}, \tag{14}$$

with equality if and only if $w(t) \equiv Ae^{-ct^2}$, where $c = 4\pi^2 \sigma_{\widehat{w}}^2$. The last result shows that.

THEOREM **D1.3.** *The Gabor window is optimal, in the sense that it minimizes the uncertainty* $\sigma_{\widehat{w}} \sigma_w$.

The windowed FT is a continuous transform, in that the local time–frequency content of a signal is contained in a function of two continuous arguments. It would be interesting to have a discrete version, that is, a decomposition of the signal along a Hilbert basis. More specifically, one asks the question: Is there a window $w(t)$ such that the family $\{\psi_{m,n}\}_{m\in\mathbb{Z},n\in\mathbb{Z}}$, where

$$\psi_{m,n}(t) = e^{2i\pi mt}w(t-n), \tag{15}$$

is an orthonormal basis of $L^2(\mathbb{R})$?

EXERCISE **D1.1.** *Show that the answer to the previous question is positive for the rectangular window* $w(t) = 1_{[0,1]}(t)$.

Although such "atomic" windowed FT bases do exist, they turn out to be very bad from the view point of time–frequency resolution, as the following result, called the Balian–Low theorem,[5], shows.

THEOREM **D1.4.** *If* $\{\psi_{m,n}\}_{m\in\mathbb{Z},n\in\mathbb{Z}}$, *where* $\psi_{m,n}$ *is defined by* (15) *(with* $g \in L^2(\mathbb{R})$*), is an orthonormal basis of* $L^2(\mathbb{R})$, *then at least one of the following equalities is true:*

$$\int_{\mathbb{R}} t^2|w(t)|^2 dt = \infty \text{ or } \int_{\mathbb{R}} \nu^2|\widehat{w}(\nu)|^2 d\nu = \infty.$$

EXERCISE **D1.2.** *Show that the system* $\{\psi_{m,n}\}_{m\in\mathbb{Z},n\in\mathbb{Z}}$, *where*

$$\psi_{m,n}(t) = e^{2i\pi mt}e^{-\alpha(t-n)^2},$$

$\alpha > 0$, *is not an orthonormal basis of* $L^2(\mathbb{R})$.

[5] R. Balian, Un principe d'incertitude fort en théorie du signal et en mécanique quantique, *CR Acad. Sci. Paris*, 292, Ser. II, 1981, 1357–1361; F. Low, Complete sets of wave packets, *A Passion for Physics—Essays in Honor of Geoffrey Chew*, 17–22, World Scientific: Singapore, 1985.

D2

The Wavelet Transform

D2·1 Time–Frequency Resolution of Wavelet Transforms

Definition of the Wavelet Transform

We mentioned in the introduction to Part D the shortcomings of the windowed Fourier transform. This chapter gives another approach to the time–frequency issue of Fourier analysis. The role played in the windowed Fourier transform by the family of functions

$$w_{\nu,b}(t) = w(t-b)e^{+2i\pi\nu t}, \quad b, \nu \in \mathbb{R},$$

is played in the wavelet transform (WT) by a family

$$\psi_{a,b}(t) = |a|^{-1/2}\psi\left(\frac{t-b}{a}\right), \quad a, b \in \mathbb{R},\ a \neq 0, \tag{16}$$

where $\psi(t)$ is called the *mother wavelet*. The function $\psi_{a,b}$ is obtained from the mother wavelet ψ by successively applying a change of time scale (accompanied by a change of amplitude scale in order to keep the energy constant) and a time shift (see Fig. D2.1).

DEFINITION **D2.1.** *The wavelet transform of the function $f \in L^2_{\mathbb{C}}(\mathbb{R})$ is the function $C_f : (\mathbb{R} - \{0\}) \times \mathbb{R} \mapsto \mathbb{C}$ defined by*

$$C_f(a,b) = \langle f, \psi_{a,b}\rangle = \int_{\mathbb{R}} f(t)\psi^*_{a,b}(t)\,\mathrm{d}t. \tag{17}$$

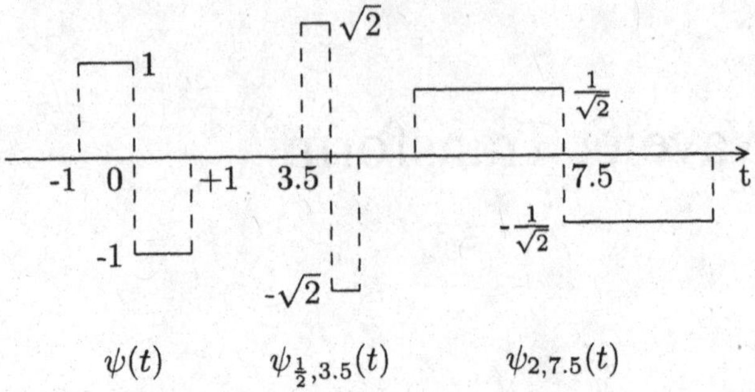

Figure D2.1. Dilations and translations

The Adaptive Time–frequency Box

By the Plancherel–Parseval identity,

$$C_f(a, b) = \langle \hat{f}, \widehat{\psi}_{a,b} \rangle = \int_{\mathbb{R}} \hat{f}(\nu)\, \widehat{\psi}^*_{a,b}(\nu)\, \mathrm{d}\nu, \tag{18}$$

where

$$\widehat{\psi}_{a,b}(\nu) = |a|^{1/2}\, e^{-2i\pi\nu b}\, \widehat{\psi}(a\nu). \tag{19}$$

Let m_ψ and σ_ψ be, respectively, the center and RMS width of the mother wavelet ψ, respectively defined by

$$m_\psi = \frac{1}{E_\psi} \int_{\mathbb{R}} t\, |\psi(t)|^2\, \mathrm{d}t,$$

$$\sigma_\psi^2 = \frac{1}{E_\psi} \int_{\mathbb{R}} (t - m_\psi)^2\, |\psi(t)|^2\, \mathrm{d}t,$$

and similarly define $m_{\widehat{\psi}}$ and $\sigma_{\widehat{\psi}}$, where $\widehat{\psi}$ is the Fourier transform of ψ. The center and width of $\psi_{a,b}$ are, respectively,

$$b + am_\psi, \qquad a\sigma_\psi,$$

whereas the center and width of $\widehat{\psi}_{a,b}$ are

$$\frac{1}{a} m_{\widehat{\psi}}, \qquad \frac{1}{a} \sigma_{\widehat{\psi}}.$$

We shall simplify notations by writing

$$m_\psi = m, \quad m_{\widehat{\psi}} = \widehat{m}, \quad \sigma_\psi = \sigma, \quad \sigma_{\widehat{\psi}} = \hat{\sigma}.$$

We see that $C_f(a, b)$ is the result of the analysis of the function f in the time–frequency box (see Fig. D2.2)

$$[b + am - a\sigma, b + am + a\sigma] \times \left[\frac{\widehat{m}}{a} - \frac{\hat{\sigma}}{a}, \frac{\widehat{m}}{a} + \frac{\hat{\sigma}}{a} \right].$$

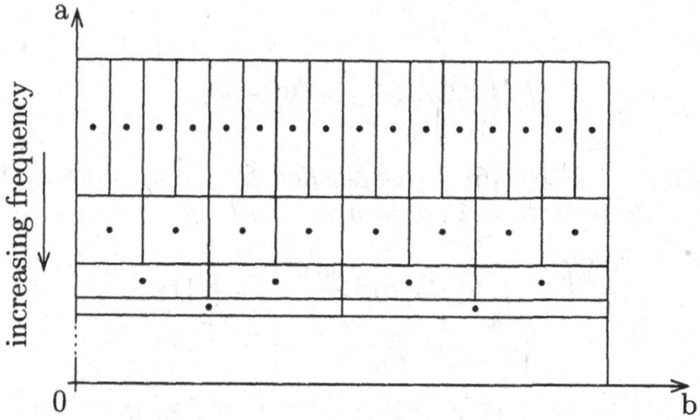

Figure D2.2. Time-frequency tiling in wavelet analysis

Let us assume that $\widehat{m} > 0$. The frequency window is then centered at $\nu = \widehat{m}/a$ and has width $2\hat{\sigma}/a$; therefore,

$$Q = \frac{\text{center frequency}}{\text{bandwidth}} = \frac{\widehat{m}}{2\hat{\sigma}}$$

is independent of the frequency variable a. This is called *constant-Q filtering*.

Calling $\nu = \widehat{m}/a$ the center frequency, we see that the area of the box is constant and equal to $4\sigma\hat{\sigma}$, but that its shape changes with the frequency $\nu = 1/a$ analyzed. For high frequencies it has a large time dimension, and for small frequencies it has a small time dimension (see Fig. D2.2). The interest of such features is discussed in the introduction to this chapter.

We shall see in the next subsection that in order to guarantee perfect reconstruction of the signal from its wavelet transform, the center of $\widehat{\psi}$ must be zero. Also, the center of the wavelet itself can be taken equal to zero without loss of generality. The Fourier transform of a wavelet has bumps at positive and negative frequencies (see Example D2.3, the Mexican hat). The centers of the bumps then play the role of the center of the wavelet in the first part of the above discussion (where $\widehat{m}$ was assumed to be nonzero).

D2·2 Wavelet Inversion Formula

Under mild conditions, there exists a wavelet inversion formula similar to the WFT inversion formula.

THEOREM **D2.1.** *Let $\psi : \mathbb{R} \mapsto \mathbb{R}$ be a mother wavelet such that $\psi \in L^1 \cap L^2$,*

$$\int_{\mathbb{R}} |\psi(t)|^2 \, dt = 1, \tag{20}$$

and

$$\int_{\mathbb{R}} \frac{|\widehat{\psi}(\nu)|^2}{|\nu|}\, d\nu = K < \infty. \tag{21}$$

Define $\psi_{a,b} \in L^2 \cap L^1$ *by (16). To the function* $f \in L^2$ *is associated its wavelet transform* $C_f : \mathbb{R}\backslash\{0\} \times \mathbb{R} \to \mathbb{C}$*, defined by (17). Then*

$$\frac{1}{K}\int_{\mathbb{R}\backslash\{0\}}\int_{\mathbb{R}} |C_f(a,b)|^2 \frac{da\, db}{a^2} = \int_{\mathbb{R}} |f(t)|^2\, dt \tag{22}$$

and

$$f(t) = \frac{1}{K}\int_{\mathbb{R}\backslash\{0\}}\int_{\mathbb{R}} C_f(a,b)\psi_{a,b}(t)\frac{da\, db}{a^2}, \tag{23}$$

in the sense that $f_\varepsilon \to f$ *in* L^2 *as* $\varepsilon \to 0^+$*, where*

$$f_\varepsilon(t) = \frac{1}{K}\iint_{\mathbb{R}\times\{|a|\geq\varepsilon\}} C_f(a,b)\psi_{a,b}(t)\frac{da\, db}{a^2}.$$

Proof: First observe that $\psi_{a,b}$ has the same energy as ψ, equal to unity. From (18) and (19),

$$C_f(a,b) = |a|^{1/2}\int \hat{f}(\nu)\widehat{\psi}^*(a\nu)e^{-2i\pi\nu b}\, d\nu$$

$$= |a|^{1/2}\mathcal{F}_\nu^{-1}\{\hat{f}(\nu)\widehat{\psi}^*(a\nu)\}(b).$$

The function inside the curly brackets is in L^1 because it is the product of two L^2-functions, and it is in L^2 as it is the product of an L^2-function with a bounded function ($\widehat{\psi}$ is bounded because $\psi \in L^1$). By the Plancherel–Parseval identity,

$$\int_{\mathbb{R}} |C_f(a,b)|^2\, db = |a|\int_{\mathbb{R}} \left|\mathcal{F}_\nu^{-1}\{\hat{f}(\nu)\widehat{\psi}^*(a\nu)\}\right|^2(b)\, db$$

$$= |a|\int_{\mathbb{R}} |\hat{f}(\nu)|^2\, |\widehat{\psi}(a\nu)|^2\, d\nu,$$

and therefore,

$$\int_{\mathbb{R}}\int_{\mathbb{R}} |C_f(a,b)|^2 \frac{db\, da}{a^2} = \int_{\mathbb{R}}\int_{\mathbb{R}} |\hat{f}(\nu)|^2\, |\widehat{\psi}(a\nu)|^2\, d\nu \frac{da}{|a|}$$

$$= \int_{\mathbb{R}} |\hat{f}(\nu)|^2 \left[\int_{\mathbb{R}} \frac{|\widehat{\psi}(a\nu)|^2}{|a|}\, da\right] d\nu$$

$$= \int_{\mathbb{R}} |\hat{f}(\nu)|^2 K\, d\nu = K\int_{\mathbb{R}} |f(t)|^2\, dt.$$

This proves (22). To prove (23), first compute

$$\begin{aligned} I(a) &= \int_{\mathbb{R}} C_f(a,b)\psi_{a,b}(t)\,\mathrm{d}b \\ &= |a|^{1/2}\int_{\mathbb{R}} \mathcal{F}_\nu^{-1}\{\hat{f}(\nu)\widehat{\psi}^*(a\nu)\}(b)\psi_{a,b}(t)\,\mathrm{d}b \\ &= |a|^{1/2}\int_{\mathbb{R}} \hat{f}(\nu)\widehat{\psi}^*(a\nu)\mathcal{F}_{\bar{b}}^{-1}\{\psi_{a,b}(t)\}(\nu)\,\mathrm{d}\nu, \end{aligned}$$

where we have used the Plancherel–Parseval identity. Now,

$$\mathcal{F}_{\bar{b}}^{-1}\{\psi_{a,b}(t)\}(\nu) = |a|^{1/2}\widehat{\psi}(a\nu)e^{2i\pi\nu t},$$

and therefore,

$$I(a) = |a|\int_{\mathbb{R}} \hat{f}(\nu)|\widehat{\psi}(a\nu)|^2 e^{2i\pi\nu t}\,\mathrm{d}\nu, \tag{24}$$

and, for $\varepsilon > 0$,

$$\begin{aligned} f_\varepsilon(t) &= \frac{1}{K}\int_{\{|a|\geq\varepsilon\}} I(a)\,\frac{\mathrm{d}a}{a^2} \\ &= \frac{1}{K}\int_{\{|a|\geq\varepsilon\}}\left(\int_{\mathbb{R}} \hat{f}(\nu)|\widehat{\psi}(a\nu)|^2 e^{2i\pi\nu t}\,\mathrm{d}\nu\right)\frac{\mathrm{d}a}{|a|}. \end{aligned} \tag{25}$$

With a view to applying Fubini's theorem we must check that the function

$$(a,\nu) \to |\hat{f}(\nu)|\,|\widehat{\psi}(a\nu)|^2\,\frac{1}{|a|}$$

is integrable in the domain $\mathbb{R}\times\{|a|\geq\varepsilon\}$. We have

$$\begin{aligned} J &= \int_{\mathbb{R}}\int_{|a|\geq\varepsilon} |\hat{f}(\nu)|\,|\widehat{\psi}(a\nu)|\,\frac{1}{|a|}\,\mathrm{d}a\,\mathrm{d}\nu \\ &= \int_{\mathbb{R}} |\hat{f}(\nu)|\left(\int_{|a|\geq\varepsilon}\frac{|\widehat{\psi}(a\nu)|^2}{|a|}\,\mathrm{d}a\right)\mathrm{d}\nu \\ &= \int_{\mathbb{R}} |\hat{f}(\nu)|\left(\int_{|x|\geq\varepsilon|\nu|}\frac{|\widehat{\psi}(x)|^2}{|x|}\,\mathrm{d}x\right)\mathrm{d}\nu \\ &= \int_{-1}^{+1} + \int_{|\nu|\geq 1} = J_1 + J_2. \end{aligned}$$

But

$$\begin{aligned} J_1 &= \int_{-1}^{+1} |\hat{f}(\nu)|\left(\int_{|x|\geq\varepsilon|\nu|}\frac{|\widehat{\psi}(x)|^2}{|x|}\,\mathrm{d}x\right)\mathrm{d}\nu \\ &\leq K\int_{-1}^{+1} |\hat{f}(\nu)|\,\mathrm{d}\nu < \infty \end{aligned}$$

because $\hat{f} \in L^2(\mathbb{R})$ and, in particular, $\hat{f} \in L^2([-1, +1])$. Also,

$$J_2 = \int_{|\nu|\geq 1} |\hat{f}(\nu)| \left(\int_{|x|\geq \varepsilon|\nu|} \frac{|\widehat{\psi}(x)|^2}{|x|} \,\mathrm{d}x \right) \mathrm{d}\nu$$

$$\leq \int_{|\nu|\geq 1} |\hat{f}(\nu)| \frac{1}{\varepsilon|\nu|} \,\mathrm{d}\nu \int_{|x|\geq \varepsilon|\nu|} |\widehat{\psi}(x)|^2 \,\mathrm{d}x$$

$$\leq \int_{\mathbb{R}} |\widehat{\psi}(x)|^2 \,\mathrm{d}x \int_{|\nu|\geq 1} |\hat{f}(\nu)| \frac{1}{\varepsilon|\nu|} \,\mathrm{d}\nu$$

$$= \frac{1}{\varepsilon} \int |\psi(t)|^2 \,\mathrm{d}t \int_{|\nu|\geq 1} \frac{|\hat{f}(\nu)|}{|\nu|} \,\mathrm{d}\nu.$$

But

$$\int_{\mathbb{R}} |\psi(t)|^2 \,\mathrm{d}t = 1,$$

and, therefore, using Schwarz' inequality,

$$J_2 \leq \frac{1}{\varepsilon} \int_{|\nu|\geq 1} \frac{|\hat{f}(\nu)|}{|\nu|} \,\mathrm{d}\nu$$

$$\leq \frac{1}{\varepsilon} \left(\int_{|\nu|\geq 1} |\hat{f}(\nu)|^2 \,\mathrm{d}\nu \right)^{1/2} \left(\int_{|\nu|\geq 1} \frac{\mathrm{d}\nu}{\nu^2} \right)^{1/2}.$$

We can therefore change the order of integration in (25):

$$Kf_\varepsilon(t) = \int_{\mathbb{R}} \hat{f}(\nu) e^{2i\pi\nu t} \left(\int_{|a|\geq \varepsilon} \frac{|\widehat{\psi}(a\nu)|^2}{|a|} \,\mathrm{d}a \right) \mathrm{d}\nu$$

$$= \int_{\mathbb{R}} \hat{f}(\nu) e^{2i\pi\nu t} g_\varepsilon(\nu) \,\mathrm{d}\nu = \mathcal{F}^{-1}\{\hat{f} g_\varepsilon\}(t).$$

Since we want to prove that $f_\varepsilon \to f$ in L^2, we must evaluate the L^2-norm of $f - f_\varepsilon$:

$$K^2 \|f - f_\varepsilon\|^2 = K^2 \|\mathcal{F}^{-1}\{f - f_\varepsilon\}\|^2 = \|\hat{f}(K - g_\varepsilon)\|^2$$

$$= \int_{\mathbb{R}} |K - g_\varepsilon(\nu)|^2 |\hat{f}(\nu)|^2 \,\mathrm{d}\nu = \int_{|\nu|\leq \varepsilon^{-1/2}} + \int_{|\nu|> \varepsilon^{-1/2}} = A + B.$$

On $\{|\nu| \leq \varepsilon^{-1/2}\}$,

$$g_\varepsilon(\nu) = \int_{|a|\geq \varepsilon} \frac{|\widehat{\psi}(a\nu)|^2}{|a|} \,\mathrm{d}a = \int_{|x|\geq \varepsilon|\nu|} \frac{|\widehat{\psi}(x)|^2}{|x|} \,\mathrm{d}x$$

$$\geq \int_{|x|\geq \varepsilon^{1/2}} \frac{|\widehat{\psi}(x)|^2}{|x|} \,\mathrm{d}x = K_\varepsilon,$$

where

$$K_\varepsilon \le K \quad \text{and} \quad \lim_{\varepsilon \to 0^+} K_\varepsilon = K.$$

Therefore,

$$A \le (K - K_\varepsilon)^2 \int_{|\nu| \le \varepsilon^{-1/2}} |\hat{f}(\nu|^2 \, \mathrm{d}\nu$$

$$\le (K - K_\varepsilon)\|f\|^2 \to 0 \quad \text{as } \varepsilon \to 0^+.$$

Also,

$$B = \int_{|\nu| > \varepsilon^{-1/2}} |\hat{f}(\nu)|^2 |K - g_\varepsilon(\nu)|^2 \, \mathrm{d}\nu$$

$$\le K^2 \int_{|\nu| > \varepsilon^{-1/2}} |\hat{f}(\nu)|^2 \, \mathrm{d}\nu \to 0 \quad \text{as } \varepsilon \to 0^+$$

since $\hat{f} \in L^2$. We have therefore proved that

$$\|f - f_\varepsilon\| \to 0 \quad \text{as } \varepsilon \to 0^+.$$

The proof is lengthy because we have only required f to be in L^2.

THEOREM **D2.2.** *If $f \in L^2 \cap L^1$ and $\hat{f} \in L^1$, the inversion formula* (23) *is true for almost all t.*

Proof: We start from (24):

$$\int_{\mathbb{R}} I(a) \frac{\mathrm{d}a}{a^2} = \int_{\mathbb{R}} \int_{\mathbb{R}} \hat{f}(\nu) |\widehat{\psi}(a\nu)|^2 e^{2i\pi\nu t} \frac{\mathrm{d}\nu \, \mathrm{d}a}{a^2}$$

$$= \int_{\mathbb{R}} \hat{f}(\nu) e^{2i\pi\nu t} \left(\int_{\mathbb{R}} \frac{|\widehat{\psi}(a\nu)|^2}{|a|} \, \mathrm{d}a \right) \mathrm{d}\nu$$

$$= K \int_{\mathbb{R}} \hat{g}(\nu) e^{2i\pi\nu t} \, \mathrm{d}\nu.$$

This quantity is almost everywhere equal to $Kf(t)$ by the Fourier inversion formula in L^1. ■

Recall that if f is continuous the equality in the Fourier inversion formula holds for all t and, therefore, (23) is then true for all $t \in \mathbb{R}$.

Oscillation Condition

Since $\widehat{\psi}$ is continuous ($\psi \in L^1$), the assumption (21) implies that $\widehat{\psi}(0) = 0$, that is, to say,

$$\int_{\mathbb{R}} \psi(t) \, \mathrm{d}t = 0. \tag{26}$$

In most situations it suffices to verify (26), and then (21) follows. For example, if $\psi(t)$ and $t\psi(t)$ are integrable, then $\widehat{\psi}$ is C^1; therefore, if $\widehat{\psi}(0) = 0$, the quantity $|\widehat{\psi}(\nu)|^2/|\nu|$ is integrable in a neighborhood of zero and therefore on $\mathbb{R}$, since at infinity there is no problem, due to the hypothesis $\psi \in L^2$ (which implies that $\widehat{\psi} \in L^2$).

EXAMPLE **D2.1** (Morlet's pseudo-wavelet). *Morlet used the mother wavelet*

$$\psi(t) = \gamma e^{-t^2/2} \cos(5t),$$

where γ is a normalization factor that makes the energy equal to unity. The theoretical problem here is that

$$\widehat{\psi}(0) = \int_{\mathbb{R}} \psi(t)\,\mathrm{d}t > 0.$$

However, the numerical results obtained with this wavelet were satisfactory because the value of $\widehat{\psi}(0)$ is in fact very small.

EXAMPLE **D2.2** (Haar wavelet). *The Haar wavelet*

$$\psi(t) = \begin{cases} 1 & \text{if } 0 \le t < \frac{1}{2}, \\ -1 & \text{if } \frac{1}{2} \le t < 1, \\ 0 & \text{otherwise}, \end{cases}$$

satisfies the conditions for the reconstruction formula (23) *to be valid. Here*

$$\widehat{\psi}(\nu) = i e^{-i\pi\nu} \frac{1 - \cos(\pi\nu)}{\pi\nu}.$$

EXAMPLE **D2.3.** *In practice, a mother wavelet ψ should be well localized in addition to $\widehat{\psi}$, and it should also be oscillating (so as to guarantee at least that $\int_{\mathbb{R}} \psi(t)\,\mathrm{d}t = 0$). Derivatives of the Gaussian pulse are good for this purpose. For example, the second-order derivative, called the* Mexican hat *(see Fig. D2.3),*

$$\psi(t) \sim (1 - t^2)e^{-t^2/2}$$

with the Fourier transform

$$\widehat{\psi}(\nu) \sim \nu^2 e^{-2\pi^2\nu^2}$$

is interesting because both ψ and $\widehat{\psi}$ are rapidly decreasing C^∞-functions.

We shall now give a pictorial example. Fig. D2.4 shows a simple signal and Fig. D2.5 shows its wavelet transform. The mother wavelet used is not given, since it is irrelevant to this qualitative illustration. In the latter figure, the time axis is horizontal, and the time axis vertical, the bottom part corresponding to high frequencies. We observe the good time localization and the fact that sharp discontinuities are represented in the bottom part.

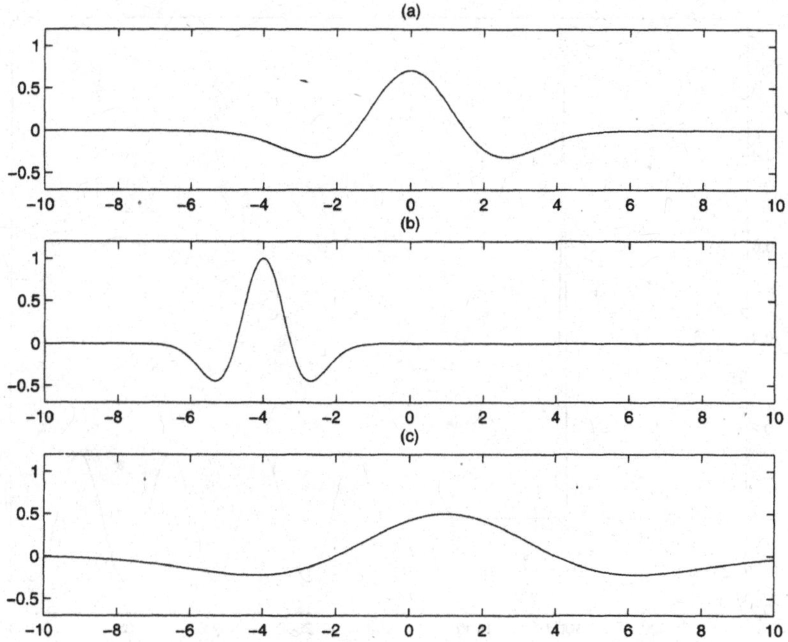

Figure D2.3. The Mexican hat

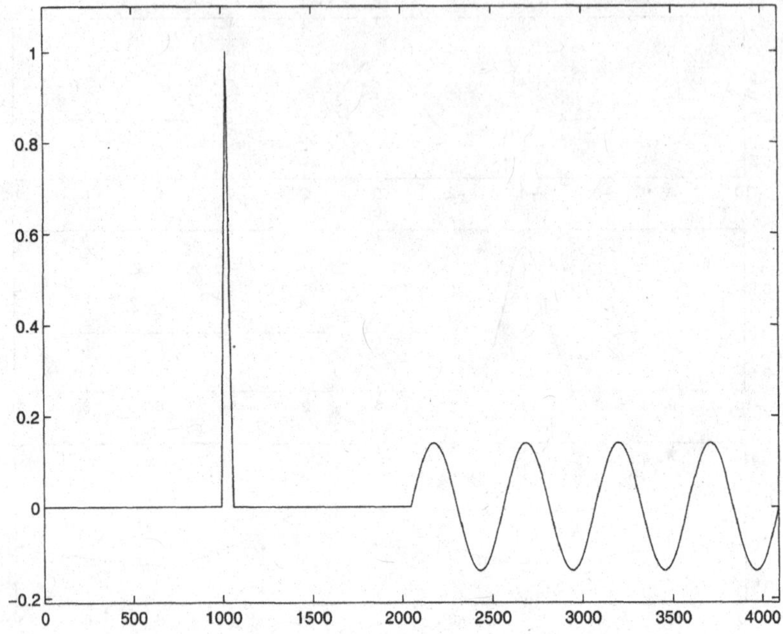

Figure D2.4. Spike + sinusoid

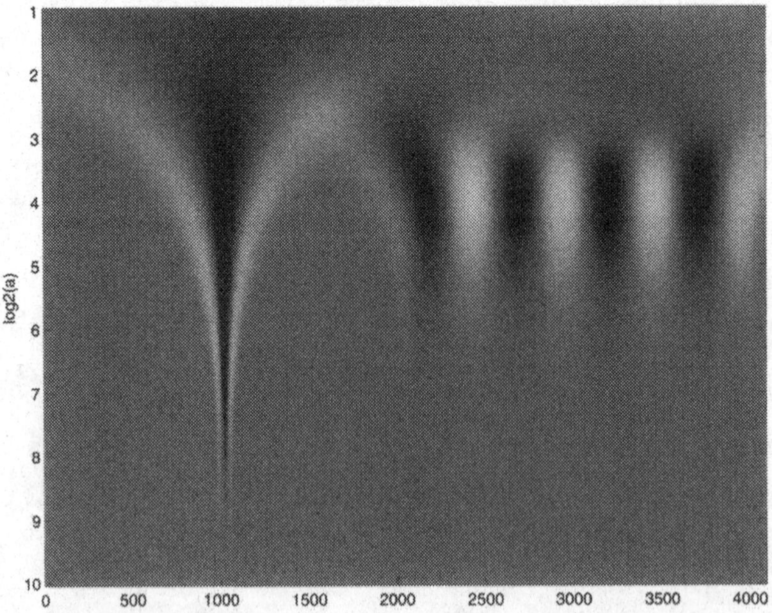

Figure D2.5. The wavelet transform of the signal in Fig. D2.4

D3
Wavelet Orthonormal Expansions

D3·1 Mother Wavelet

The wavelet analysis of Chapter D2 is *continuous*, in that the original function of L^2 is reconstructed as an integral, not as a sum. One would rather store the original function not as a function of two arguments, but as the doubly indexed sequence of coefficients of a decomposition along an orthonormal base of L^2. Multiresolution analysis is one particular way of obtaining such orthonormal bases.

In the remainder, we adopt a slightly different definition of the Fourier transform. The FT $\hat{f}(\omega)$ of the signal $f(t)$ is now defined by

$$\hat{f}(\omega) = \int_{\mathbb{R}} f(t)e^{-i\omega t}\, \mathrm{d}t.$$

The inversion formula, when it holds true, takes the form

$$f(t) = \frac{1}{2\pi}\int_{\mathbb{R}} \hat{f}(\omega)e^{i\omega t}\, \mathrm{d}\omega, \text{ a.e.,}$$

and the Plancherel–Parseval identity, when it holds true, reads

$$\int_{\mathbb{R}} f(t)g(t)^*\, \mathrm{d}t = \frac{1}{2\pi}\int_{\mathbb{R}} \hat{f}(\omega)\hat{g}(\omega)^*\, \mathrm{d}\omega.$$

Also, the necessary and sufficient condition for $\{\varphi(\cdot - n)\}_{n\in\mathbb{Z}}$ to be an orthonormal sequence of $L^2_{\mathbb{C}}(\mathbb{R})$ (Theorem C4.5) now reads

$$\sum_{k\in\mathbb{Z}} |\hat{\varphi}(\omega + 2k\pi)|^2 = 1, \text{ a.e.}$$

One reason for abandoning the definition in terms of the frequency ν is that the topic of MRA involves a mixture of analog signals and of digital filtering, and digital signal processing is traditionally—as in the present text—dealt with in terms of the pulsation ω.

Scaling Function and MRA

DEFINITION **D3.1.** *A multiresolution analysis (MRA) of* $L^2 = L^2_{\mathbb{C}}(\mathbb{R})$ *consists of a function* $\varphi \in L^2$ *together with a family* $\{V_j\}_{j\in\mathbb{Z}}$ *of Hilbert subspaces of* L^2 *such that*

(a) $\{\varphi(\cdot - n)\}_{n\in\mathbb{Z}}$ *is an orthonormal basis of* V_0,

(b) for all $j \in \mathbb{Z}$, $V_j \subseteq V_{j+1}$ *(the* V_j*'s are said to be nested; see Fig. 4.7),*

(c) $f \in V_0 \Longleftrightarrow f(2^j \cdot) \in V_j$,

(d) $\bigcap_{j\in\mathbb{Z}} V_j = \varnothing$ *and* $\operatorname{clos}\left(\bigcup_{j\in\mathbb{Z}} V_j\right) = L^2$.

The function φ is called the *scaling function* of the MRA. The index j represents a resolution level: The projection $P_j f$ of a function $f \in L^2$ on V_j is interpreted as the observation of this function at the *resolution level* j.

Usually, the projection on V_0 is the function itself, in which case the projections at all levels $j \geq 0$ are identical. The projection at level 0 is, in applications, the one offered by the recording device.

Observe that, since the mapping $f \to \sqrt{2}\, f(2\,\cdot)$ is an isometry from V_0 onto V_1 and since $\{\varphi(\cdot - n)\}_{n\in\mathbb{Z}}$ is an orthonormal basis of V_0, the set $\{\sqrt{2}\,\varphi(2\,\cdot - n)\}_{n\in\mathbb{Z}}$ is an orthonormal basis of V_1. More generally, $\{\varphi_{j,n}\}_{n\in\mathbb{Z}}$ is an orthonormal basis of V_j, where

$$\varphi_{j,n}(t) := 2^{j/2}\varphi(2^j t - n). \tag{27}$$

With respect to (d), recall that (Exercise C1.6)

$$\bigcap_{j\in\mathbb{Z}} V_j = \varnothing \Longleftrightarrow \lim_{j\to+\infty} P_{-j} f = 0 \quad \text{for all } f \in L^2 \tag{28}$$

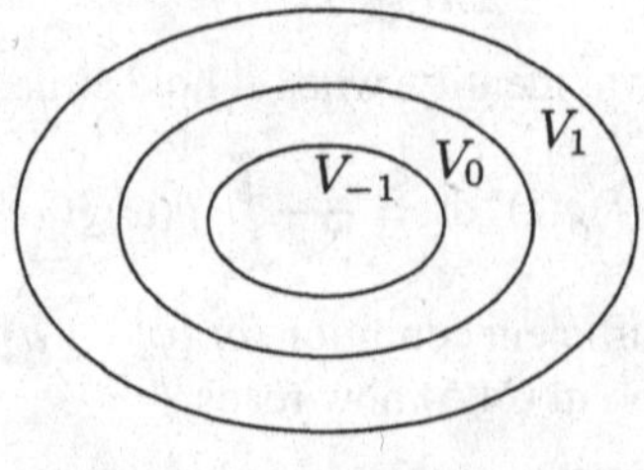

Figure D3.1. Nested subspaces

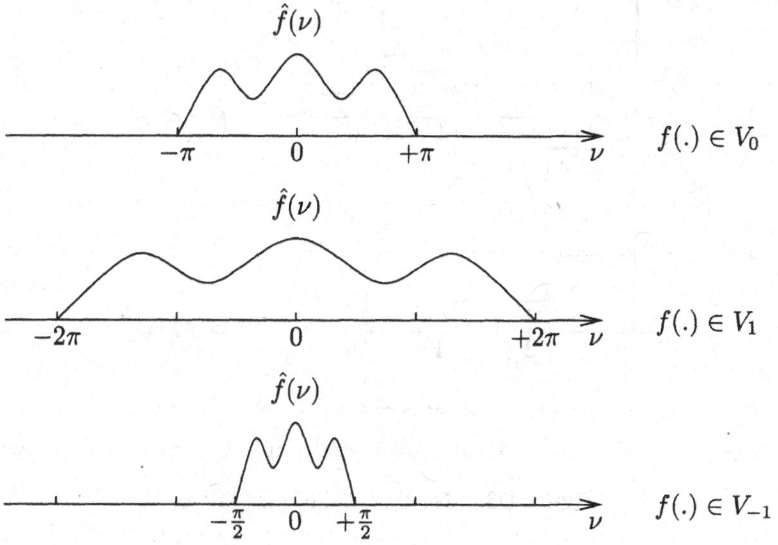

Shannon multiresolution

Figure D3.2. Nesting in the Shannon MRA

and

$$\text{clos} \bigcup_{j \in \mathbb{Z}} V_j = L^2 \iff \lim_{j \to +\infty} P_j f = f \quad \text{for all } f \in L^2. \tag{29}$$

EXERCISE **D3.1** **(Shannon scaling function).** *For all $j \in \mathbb{Z}$, define*

$$V_j = \{f \in L^2 : \text{supp}\,(\hat{f}) \subset [-2^j\pi, +2^j\pi]\}.$$

Define the function φ by its Fourier transform

$$\widehat{\varphi}(\omega) = \mathbb{1}_{[-\pi,+\pi]}(\omega)$$

(see Fig. D3.2). Verify that $\{V_j\}_{j\in\mathbb{Z}}$ is a multiresolution analysis of L^2 associated with the scaling function φ.

EXERCISE **D3.2** **(Haar scaling function).** *For all $j \in \mathbb{Z}$, define*

$$V_j = \{f \in L^2 : f \text{ is a.e. constant on the intervals } (k2^{-j},\ (k+1)2^{-j}]\}.$$

Define

$$\varphi(t) = \mathbb{1}_{(0,1]}(t)$$

(see Fig. D3.3). Verify that $\{V_j\}_{j\in\mathbb{Z}}$ is a multiresolution analysis of L^2 associated with the scaling function φ.

We shall see later that some regularity of the scaling function is desirable.

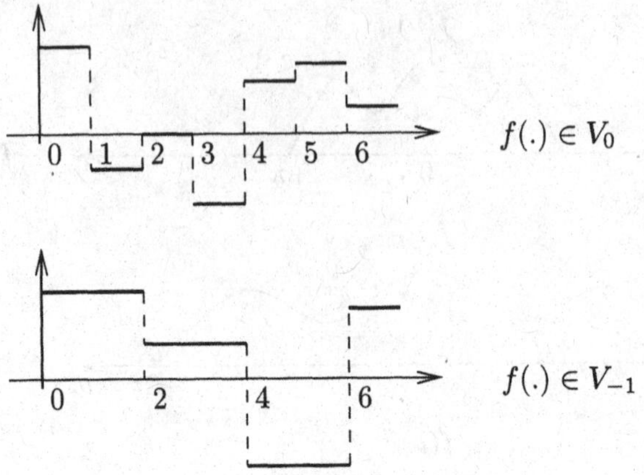

Haar multiresolution

Figure D3.3. Nesting in the Haar MRA

DEFINITION **D3.2.** *The function φ is said to belong to $\mathcal{S}_r$ for some $r \in \mathbb{N}$ if φ is r times continuously differentiable with rapidly decreasing derivatives, in the sense that*

$$|\varphi^{(k)}(t)| \le \frac{C_{kp}}{(1+|t|)^p}, \qquad \text{for } k = 0, 1, \dots, r, \quad \text{and all } p \in \mathbb{N}. \tag{30}$$

This is a multiresolution analysis for which the scaling function $\varphi \in S^r$ is called r-smooth.

The Haar and Shannon scaling functions are not in $\mathcal{S}_r$ (for any $r \in \mathbb{N}$).

Conditions (a), (b), (c), and (d) of Definition 4.3 are not independent. In fact, the first part of (d) is always true under conditions (a), (b), (c), whereas the latter conditions are almost sufficient for the second part of (d). The result below makes this statement precise.

THEOREM **D3.1.** *Suppose that conditions (a), (b), and (c) of the definition of an MRA are satisfied. Then $\bigcap_{j\in\mathbb{Z}} V_j = \varnothing$. Moreover, if $\widehat{\varphi}$ is continuous at the origin, then*

$$|\widehat{\varphi}(0)| \neq 0 \Longleftrightarrow \operatorname{clos}\bigcup_{j\in\mathbb{Z}} V_j = L^2_{\mathbb{C}}(\mathbb{R}). \tag{31}$$

In this case, necessarily, $|\widehat{\varphi}(0)| = 1$.

Proof: The first statement will be proven in the more general Theorem D4.1.

We now prove the second statement. (The proof is technical and can be skipped in a first reading.) Denote by T_a the translation operator defined by $T_a f(x) = f(x-a)$. We shall first show that the Hilbert space $W = \operatorname{clos}\left(\bigcup_{j\in\mathbb{Z}} V_j\right)$ is invariant under translations.

We begin with dyadic translations ($a = m2^{-\ell}$, where $\ell, m \in \mathbb{Z}$). Let $f \in W$. Therefore, for any given $\varepsilon > 0$, there exist $j_0 \in \mathbb{Z}$ and $h \in V_{j_0}$ such that $\|f - h\|_2 \le \varepsilon$. For all $j \ge j_0$, h is also in V_j. In particular,

$$h = \sum_{k \in \mathbb{Z}} c_k^j \varphi(2^j \cdot -k),$$

and the function

$$T_{m2^{-\ell}} h = \sum_{k \in \mathbb{Z}} c_k^j \varphi(2^j \cdot -2^{j-\ell} m - k)$$

is therefore in V_j if $j \ge \ell$ (because $2^{j-\ell}m$ is then in $\mathbb{Z}$). Therefore, for all $j \ge \ell$,

$$\|T_{m2^{-\ell}} f - T_{m2^{-\ell}} h\|_2 = \|f - h\|_2 \le \varepsilon$$

and $T_{m2^{-\ell}} h \in V_j$. This means that $T_{m2^{-\ell}} f$ is ε-close to V_j for all $j \ge \ell$. From this and the arbitrariness of ε, we deduce that $T_{m2^{-\ell}} f \in W$.

Let now $a \in \mathbb{R}$ be arbitrary. Given $\varepsilon > 0$, there exists δ such that, for all $c \in (a - \delta, a + \delta)$,

$$\|T_a f - T_c f\|_2 < \varepsilon$$

(use Theorem C3.1 stating that the map $a \mapsto T_a f$ is uniformly continuous). In particular, we can find a dyadic number c for which the above inequality is satisfied. Since $T_c f \in W$ and ε is arbitrary, we deduce that $T_a f \in W$.

We now proceed to the proof of (31). We assume that $\widehat{\varphi}$ is continuous at 0 and that $|\widehat{\varphi}(0)| \neq 0$. Therefore, $\widehat{\varphi}(\omega) \neq 0$ on $(-c, +c)$, for some $c > 0$. Consider any function g orthogonal to W, that is, orthogonal to all $f \in W$. Since W is invariant under translations, for all $x \in \mathbb{R}$ and for all $f \in W$,

$$0 = \int_{\mathbb{R}} f(x + t) g(t)^* \, dt.$$

By the Plancherel–Parseval identity,

$$0 = \int_{\mathbb{R}} e^{i\omega x} \hat{f}(\omega) \hat{g}(\omega)^* \, d\omega,$$

for all $x \in \mathbb{R}$. The function $\hat{f}\hat{g}^* \in L^1_{\mathbb{C}}(\mathbb{R})$, and therefore, by the Fourier inversion theorem in L^1, $\hat{f}\hat{g}^* = 0$ almost everywhere.

In particular, with $f(t) = 2^j \varphi(2^j t)$ (indeed, such $f \in V_j \subset W$), we obtain

$$\widehat{\varphi}(2^{-j}\omega)\hat{g}(\omega)^* = 0, \quad \text{a.e.}$$

Since $\widehat{\varphi}(2^{-j}\omega) \neq 0$ if $\omega \in (-2^j c, +2^j c)$, we have that $\hat{g}(\omega) = 0$ if $\omega \in (-2^j c, +2^j c)$. This being true for all $j \in \mathbb{Z}$, we have that $\hat{g}$, and therefore g, is almost everywhere null. We have thus proven that the only function in $L^2_{\mathbb{C}}(\mathbb{R})$ orthogonal to W is the null function. Therefore, W is exactly $L^2_{\mathbb{C}}(\mathbb{R})$.

Assume now that $W = L^2_{\mathbb{C}}(\mathbb{R})$. Let f be the function with the FT $\hat{f} = 1_{[-1,+1]}$. In particular,

$$\| f \|_2^2 = \frac{1}{2\pi} \| \hat{f} \|_2^2 = \frac{1}{\pi}.$$

By (29),

$$\lim_{j \uparrow +\infty} \| f - P_j f \|_2 = 0,$$

and therefore, by continuity of the norm,

$$\lim_{j \uparrow +\infty} \| P_j f \|_2^2 = \| f \|_2^2 = \frac{1}{\pi},$$

that is,

$$\lim_{j \uparrow +\infty} \| \sum_{k \in \mathbb{Z}} \langle f, \varphi_{j,k} \rangle \varphi_{j,k} \|_2^2 = \frac{1}{\pi}.$$

We have by the Plancherel–Paseval identity for orthonormal bases (Theorem C2.2),

$$\| \sum_{k \in \mathbb{Z}} \langle f, \varphi_{j,k} \rangle \varphi_{j,k} \|_2^2 = \sum_{k \in \mathbb{Z}} \left| \int_{\mathbb{R}} f(t) \varphi_{j,-k}(t)^* \, \mathrm{d}t \right|^2 .$$

The last sum equals, by another Plancherel–Paseval identity,

$$\begin{aligned}
&\frac{1}{4\pi^2} \sum_{k \in \mathbb{Z}} \left| \int_{\mathbb{R}} \hat{f}(\omega) \widehat{\varphi}_{j,-k}(\omega)^* \mathrm{d}\omega \right|^2 \\
&\quad = \frac{1}{4\pi^2} \sum_{k \in \mathbb{Z}} \left| \int_{\mathbb{R}} \hat{f}(\omega) e^{-i2^{-j}k\omega} \widehat{\varphi}(2^{-j}\omega)^* \mathrm{d}\omega \right|^2 \\
&\quad = 2^j \sum_{k \in \mathbb{Z}} \left| \frac{1}{2\pi} \int_{-2^{-j}}^{+2^{-j}} e^{-ik\omega} \widehat{\varphi}(\omega)^* \mathrm{d}\omega \right|^2 .
\end{aligned}$$

For large enough j, $[-2^{-j}, +2^{-j}] \subset [-\pi, +\pi]$, and therefore the last displayed expression is 2^j times the sum of the squared absolute values of the Fourier coefficients of $1_{[-2^{-j},+2^{-j}]}\widehat{\varphi}^*$. Therefore, by the appropriate Plancherel–Parseval identity,

$$\sum_{k \in \mathbb{Z}} \left| \frac{1}{2\pi} \int_{-2^{-j}}^{+2^{-j}} e^{-ik\omega} \widehat{\varphi}(\omega)^* \, \mathrm{d}\omega \right|^2 = \frac{1}{2\pi} \int_{-2^{-j}}^{+2^{-j}} |\widehat{\varphi}(\omega)|^2 \, \mathrm{d}\omega = \frac{1}{\pi}.$$

Therefore,

$$\lim_{j \uparrow +\infty} \frac{2^j}{2\pi} \int_{-2^{-j}}^{+2^{-j}} |\widehat{\varphi}(\omega)|^2 \, \mathrm{d}\omega = \frac{1}{\pi}.$$

By continuity of $\widehat{\varphi}$ at 0, this limit is also $\frac{1}{\pi} |\widehat{\varphi}(0)|^2$. Therefore, $|\widehat{\varphi}(0)| = 1$. ■

Wavelet Expansion

We shall suppose in the sequel that the scaling functions φ satisfy $|\widehat{\varphi}(0)| > 0$, and then take (without further loss of generality)

$$\widehat{\varphi}(0) = 1. \tag{32}$$

DEFINITION **D3.3.** *A wavelet orthonormal basis of* $L^2 = L^2_{\mathbb{C}}(\mathbb{R})$ *is an orthonormal basis of the form* $\{\psi_{j,n}\}_{j,n\in\mathbb{Z}}$, *where*

$$\psi_{j,n}(t) = 2^{\frac{j}{2}}\psi(2^j t - n). \tag{33}$$

The function ψ is then called the *mother wavelet* of the wavelet basis. The expansion

$$f = \sum_{j\in\mathbb{Z}}\sum_{k\in\mathbb{Z}}\langle f, \psi_{j,k}\rangle\psi_{j,k} \tag{34}$$

is called the *wavelet expansion* of f.

A wavelet orthonormal basis can be obtained from an MRA in the following way. Let W_j be the orthogonal complement of V_j in V_{j+1}:

$$V_{j+1} = V_j \oplus W_j. \tag{35}$$

From property (d) of the definition of MRA,

$$L^2 = \bigoplus_{j\in\mathbb{Z}} W_j. \tag{36}$$

Also, from (c),

$$f \in W_0 \Longleftrightarrow f(2^j\,\cdot) \in W_j.$$

Therefore, if we can exhibit an orthonormal basis of W_0 of the form $\{\psi(\cdot - n)\}_{n\in\mathbb{Z}}$, then $\{\psi_{j,n}\}_{n\in\mathbb{Z}}$ is an orthonormal basis of W_j. Therefore, by (36) $\{\psi_{j,n}\}_{j,n\in\mathbb{Z}}$ is an orthonormal basis of L^2.

Recall that P_j is the projection on V_j. We have

$$P_{j+1}f = P_j f + \sum_{k\in\mathbb{Z}}\langle f, \psi_{j,k}\rangle\psi_{j,k} \quad \text{for all } f \in L^2. \tag{37}$$

$P_j f$ is the result of observing f at the resolution level j: As j increases, the resolution increases (note that $V_j \subset V_{j+1}$); the difference

$$P_{j+1}f - P_j f = \sum_{k\in\mathbb{Z}}\langle f, \psi_{j,k}\rangle\psi_{j,k}$$

is the additional detail required to pass from the resolution level j to the higher resolution level $j+1$.

A first issue is: How to compute the mother wavelet ψ from the scaling function φ? The next question is: How to obtain a scaling function φ? Finally, one would like to obtain a mother wavelet with "good" numerical properties, that is fast convergence of the wavelet expansion (34).

D3·2 Mother Wavelet in the Fourier Domain

We address the first issue, that of explicitly finding a mother wavelet given a scaling function.

EXAMPLE **D3.1.** *We seek to obtain the mother wavelet corresponding to the Haar scaling function. Recall that V_1 is the Hilbert space of L^2-functions that are constant almost everywhere on the intervals $(n/2, (n+1)/2]$. The mother wavelet ψ must be of this type since $W_0 \subset V_1$. The function of norm* 1, *with support* (0, 1],

$$\psi(t) = \begin{cases} +1 & \text{if } t \in (0, \frac{1}{2}], \\ -1 & \text{if } t \in (\frac{1}{2}, 1], \end{cases}$$

does it. To see this, it suffices to verify that any $f \in V_1$ with support (0, 1] *is a linear combination of φ and ψ and that φ and ψ are orthogonal. Orthogonality is obvious. Any $f \in V_1$ such that* supp$(f) \in (0, 1]$ *is of the form*

$$f(t) = \begin{cases} \alpha & \text{if } t \in (0, \frac{1}{2}], \\ \beta & \text{if } t \in (\frac{1}{2}, 1]. \end{cases}$$

and we therefore have the decomposition (see Fig. D3.4)

$$f = \frac{\alpha + \beta}{2}\varphi + \frac{\alpha - \beta}{2}\psi.$$

The function ψ is called the Haar mother wavelet.

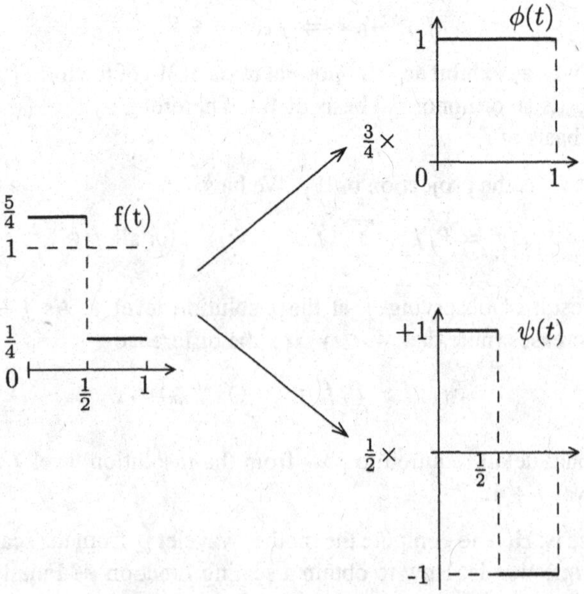

Figure D3.4. Haar decomposition

We now give a simple example of wavelet analysis to illustrate the notions of projection and detail. Fig. D3.5a gives (from top to bottom) a signal and its successive projections on the nested subspaces at decreasing resolution levels, whereas Fig. D3.5b gives (from top to bottom) the successive projections on the detail subspaces at decreasing resolution levels. In particular, the second function in Fig. D3.5b is the difference between the first and second functions in Fig. D3.5a.

The general case will now be treated. We shall obtain a necessary condition for the scaling function φ to be a scaling function. First, since $\{\varphi(\cdot - n)\}_{n\in\mathbb{Z}}$ is an orthonormal system, we have, by Theorem C4.5,

$$\sum_{k\in\mathbb{Z}} |\widehat{\varphi}(\omega + 2k\pi)|^2 = 1, \quad \text{a.e.} \tag{38}$$

The scaling function $\varphi \in V_0$ and therefore, $\varphi \in V_1$. Requirements (a) and (c) in the definition of an MRA imply that $\{\varphi_{1,n}\}_{n\in\mathbb{Z}}$ is a Hilbert basis of V_1, and we therefore have the expansion $\varphi = \sum_{n\in\mathbb{Z}} h_n \varphi_{1,n}$, that is,

$$\varphi = \sqrt{2} \sum_{n\in\mathbb{Z}} h_n \varphi(2\,\cdot\, - n), \tag{39}$$

where

$$h_n = \langle \varphi, \varphi_{1,n} \rangle. \tag{40}$$

In the Fourier domain (39) reads

$$\widehat{\varphi}(\omega) = \frac{1}{\sqrt{2}} \sum_{n\in\mathbb{Z}} h_n e^{-in\frac{\omega}{2}} \widehat{\varphi}\left(\frac{\omega}{2}\right),$$

that is,

$$\widehat{\varphi}(\omega) = m_0\left(\frac{\omega}{2}\right) \widehat{\varphi}\left(\frac{\omega}{2}\right), \tag{41}$$

where $m_0(\omega)$ is the 2π-periodic function defined by

$$m_0(\omega) = \frac{1}{\sqrt{2}} \sum_{n\in\mathbb{Z}} h_n e^{-in\omega}. \tag{42}$$

It is called the *low-pass filter* MRA, because $m_0(0) = 1$ (recall the running assumption that $\widehat{\varphi}(0) = 1$; see (32)). Substituting identity (41) in (38) gives

$$\begin{aligned} 1 &= \sum_{k\in\mathbb{Z}} \left| m_0\left(\frac{\omega}{2} + k\pi\right)\right|^2 \left|\widehat{\varphi}\left(\frac{\omega}{2} + k\pi\right)\right|^2 \\ &= \sum_{k\in\mathbb{Z}} \left| m_0\left(\frac{\omega}{2} + 2k\pi\right)\right|^2 \left|\widehat{\varphi}\left(\frac{\omega}{2} + 2k\pi\right)\right|^2 \\ &\quad + \sum_{k\in\mathbb{Z}} \left| m_0\left(\frac{\omega}{2} + \pi + 2k\pi\right)\right|^2 \left|\widehat{\varphi}\left(\frac{\omega}{2} + \pi + 2\pi\right)\right|^2 \\ &= \left| m_0\left(\frac{\omega}{2}\right)\right|^2 \sum_{k\in\mathbb{Z}} \left|\widehat{\varphi}\left(\frac{\omega}{2} + 2k\pi\right)\right|^2 \end{aligned}$$

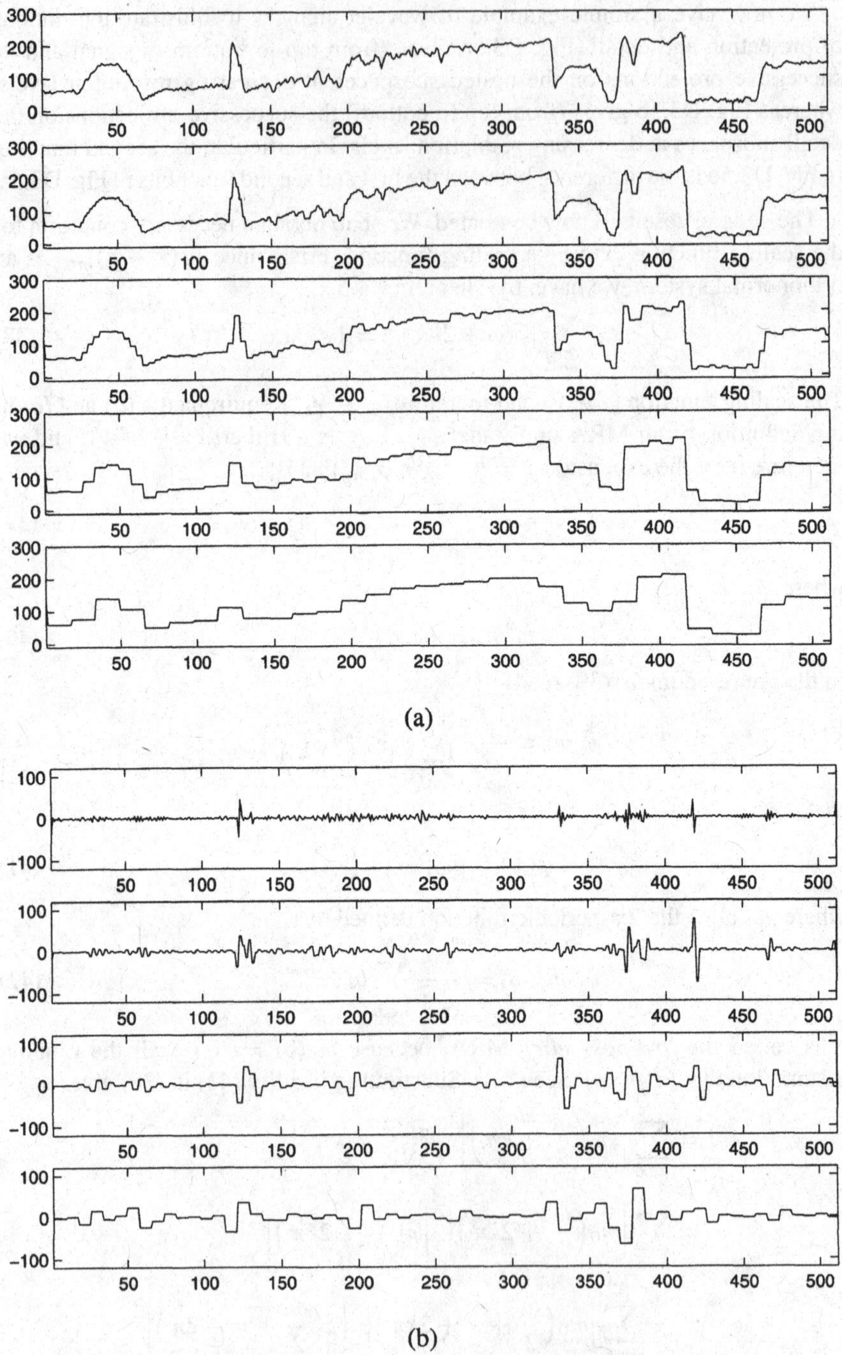

Figure D3.5. Haar wavelet analysis

$$+\left|m_0\left(\frac{\omega}{2}+\pi\right)\right|^2\sum_{k\in\mathbb{Z}}\left|\widehat{\varphi}\left(\frac{\omega}{2}+\pi+2k\pi\right)\right|^2$$

$$=\left|m_0\left(\frac{\omega}{2}\right)\right|^2+\left|m_0\left(\frac{\omega}{2}+\pi\right)\right|^2.$$

Therefore,

$$\left|m_0\left(\frac{\omega}{2}\right)\right|^2+\left|m_0\left(\frac{\omega}{2}+\pi\right)\right|^2=1,\quad \text{a.e.},$$

or, equivalently,

$$|m_0(\omega)|^2+|m_0(\omega+\pi)|^2=1,\ \text{a.e.} \tag{43}$$

The filter with frequency response $e^{i\omega}m_0(\omega+\pi)^*$ is called the *high-pass filter* of the MRA. Eqn. (43) shows that the high-pass and the low-pass filters altogether extract the whole energy contained in the band $[-\pi,+\pi]$.

We now characterize the spaces V_{-1} and V_0. This will be a preliminary to the characterization of W_{-1}, the orthogonal complement of V_{-1} in V_0. Once this is done, we shall obtain the characterization of W_0 and then the mother wavelet itself.

Lemma **D3.1.** *$f\in V_{-1}$ if and only if it has an FT of the form*

$$\hat{f}(\omega)=m(2\omega)m_0(\omega)\widehat{\varphi}(\omega), \tag{44}$$

for some 2π-periodic function $m\in L^2_{\mathbb{C}}([-\pi,+\pi])$.

Proof: Indeed, any $f\in V_{-1}$ can be decomposed along the orthonormal basis $\{\varphi_{-1,n}\}_{n\in\mathbb{Z}}$, that is,

$$f(t)=\frac{1}{\sqrt{2}}\sum_{k\in\mathbb{Z}}c_k\varphi(\frac{1}{2}t-k), \tag{45}$$

where $\{c_n\}_{n\in\mathbb{Z}}\in\ell^2_{\mathbb{C}}$. Taking the FT, we obtain

$$\hat{f}(\omega)=\sqrt{2}\sum_{k\in\mathbb{Z}}c_ke^{-i2k\omega}\widehat{\varphi}(2\omega).$$

This is (44) (using (41) and defining $m(\omega)=\sqrt{2}\sum_{k\in\mathbb{Z}}c_ke^{-ik\omega}$).

Conversely, consider a function $\hat{f}$ defined by (44), where m is a 2π-periodic function in $L^2_{\mathbb{C}}([-\pi,+\pi])$. We show that $\hat{f}$ is in $L^2_{\mathbb{C}}(\mathbb{R})$. First, observe that it is of the form $h(\omega)\widehat{\varphi}(\omega)$, where h is a 2π-periodic function in $L^2_{\mathbb{C}}([-\pi,+\pi])$ (since $m\in L^2_{\mathbb{C}}([-\pi,+\pi])$ and since m_0 is bounded in view of Eq. (43)). Now

$$\int_{\mathbb{R}}|h(\omega)\widehat{\varphi}(\omega)|^2\,\mathrm{d}\omega=\sum_{k\in\mathbb{Z}}\int_{-\pi}^{+\pi}|h(\omega)|^2|\widehat{\varphi}(\omega+2k\pi)|^2\,\mathrm{d}\omega$$

$$=\int_{-\pi}^{+\pi}|h(\omega)|^2\,\mathrm{d}\omega<+\infty.$$

This proves that $\hat{f} \in L^2_{\mathbb{C}}(\mathbb{R})$. Since $\hat{f} \in L^2_{\mathbb{C}}(\mathbb{R})$, it is the FT of a function $f \in L^2_{\mathbb{C}}(\mathbb{R})$. Tracing back the computations in the first part of the proof, we obtain that (45) holds true, with $\{c_n\}_{n\in\mathbb{Z}} \in \ell^2_{\mathbb{C}}$, which implies that $f \in V_{-1}$. ■

LEMMA **D3.2.** *$f \in V_0$ if and only if it has an FT of the form*

$$\hat{f}(\omega) = d(\omega)\widehat{\varphi}(\omega), \tag{46}$$

for some 2π-periodic function $d \in L^2_{\mathbb{C}}([-\pi, +\pi])$.

Proof: Indeed, let $f \in V_0$. It can be decomposed along the orthonormal basis $\{\varphi_{0,n}\}_{n\in\mathbb{Z}}$, that is,

$$f(t) = \sum_{k\in\mathbb{Z}} d_k \varphi(t-k), \tag{47}$$

where $\{d_n\}_{n\in\mathbb{Z}} \in \ell^2_{\mathbb{C}}$. Taking the FT, we obtain

$$\hat{f}(\omega) = \sum_{k\in\mathbb{Z}} d_k e^{ik\omega}\widehat{\varphi}(\omega) = d(\omega)\widehat{\varphi}(\omega),$$

where $d \in L^2_{\mathbb{C}}([-\pi, +\pi])$. Arguing as in the proof of Lemma D3.1, we can show that any function $\hat{f}$ of the form (46) is the FT of a function $f \in V_0$. ■

Consider the mapping $U : V_0 \mapsto L^2_{\mathbb{C}}([-\pi, +\pi])$ defined by $Uf = d$ (where d is defined by (46)). Clearly, this mapping is linear, and

$$\|Uf\|^2_{L^2_{\mathbb{C}}([-\pi,+\pi])} = \|d\|^2_{L^2_{\mathbb{C}}([-\pi,+\pi])} = 2\pi \sum_{k\in\mathbb{Z}} |d_k|^2 = 2\pi \|f\|^2_2. \tag{48}$$

By the polarization identity, for all $f, g \in V_0$,

$$\langle f, g\rangle_{L^2_{\mathbb{C}}(\mathbb{R})} = \frac{1}{2\pi}\langle Uf, Ug\rangle_{L^2_{\mathbb{C}}([-\pi,+\pi])}. \tag{49}$$

We are now ready to state and prove the Fourier characterization of W_0, the Hilbert space of details at level 0.

THEOREM **D3.2.** *The function $f \in W_0$ if and only if*

$$\hat{f}(\omega) = e^{i\frac{\omega}{2}} m_0\left(\frac{\omega}{2} + \pi\right)^* \nu(\omega)\widehat{\varphi}\left(\frac{\omega}{2}\right), \tag{50}$$

for some 2π-periodic function ν in $L^2_{\mathbb{C}}(-\pi, +\pi)$.

Proof: Observe that it is equivalent to prove that the function $f \in W_{-1}$ if and only if

$$\hat{f}(\omega) = e^{i\omega} m_0(\omega + \pi)^* \nu(2\omega)\widehat{\varphi}(\omega), \tag{51}$$

for some 2π-periodic function ν in $L^2_{\mathbb{C}}(-\pi, +\pi)$.

Let $f \in W_{-1}$, that is, $f \in V_0$ and $f \perp V_{-1}$. Being in V_0, f has a representation of type (46). By (49) and the characterization (44) of V_{-1}, the orthogonality of f and V_{-1} is equivalent to

$$0 = \int_0^{2\pi} d(\omega) m(2\omega)^* m_0(\omega)^* \, d\omega,$$

for all 2π-periodic function $m \in L^2_{\mathbb{C}}([-\pi, +\pi])$. This can also be written

$$0 = \int_0^{\pi} m(2\omega)^* \left[d(\omega)m_0(\omega)^* + d(\omega+\pi)m_0(\omega+\pi)^* \right] d\omega.$$

The function in the square brackets is therefore orthogonal to all $g \in L^2_{\mathbb{C}}([0, +\pi])$, and therefore,

$$d(\omega)m_0(\omega)^* + d(\omega+\pi)m_0(\omega+\pi)^* = 0 \tag{52}$$

almost everywhere in $[0, +\pi]$ (and therefore almost everywhere in $[-\pi, +\pi]$).

Define

$$\mathbf{m}_0(\omega) = (m_0(\omega), m_0(\omega+\pi)).$$

In view of the identity (43), this is a unitary vector in $\mathbb{C}^2$ considered as a 2-dimensional vector space (with scalar field $\mathbb{C}$). The vector

$$\mathbf{m'}_0(\omega) = (m_0(\omega+\pi)^*, -m_0(\omega)^*)$$

is unitary and orthogonal to $\mathbf{m}_0(\omega)$. Defining

$$\mathbf{d}_0(\omega) = (d(\omega), d(\omega+\pi)),$$

we have, by (52), $\mathbf{d}_0 \perp \mathbf{m}_0(\omega)$. Therefore,

$$\mathbf{d}_0 = \lambda(\omega)\mathbf{m'}_0(\omega),$$

where

$$\lambda(\omega) = \langle \mathbf{d}_0, \mathbf{m'}_0(\omega) \rangle = d(\omega)m_0(\omega+\pi) - d(\omega+\pi)m_0(\omega).$$

In particular,

$$\lambda(\omega+\pi) = -\lambda(\omega+2\pi), \text{ a.e.}$$

or, equivalently,

$$\lambda(\omega) = -\lambda(\omega+\pi), \text{ a.e.},$$

which implies in particular that λ is 2π-periodic. It is also in $L^2_{\mathbb{C}}([-\pi, +\pi])$. Indeed,

$$\hat{f}(\omega) = d(\omega)\widehat{\varphi}(\omega) = \lambda(\omega)m_0(\omega+\pi)^*\widehat{\varphi}(\omega),$$

and therefore,

$$\begin{aligned}\int_0^{\pi} |\lambda(\omega)|^2 \, d\omega &= \int_0^{\pi} |\lambda(\omega)|^2 (|m_0(\omega)|^2 + |m_0(\omega+\pi)|^2) \, d\omega \\ &= \int_0^{2\pi} |\lambda(\omega)|^2 \, |m_0(\omega+\pi)|^2 \, d\omega \\ &= \int_0^{2\pi} |d(\omega)|^2 = \sum_{n\in\mathbb{Z}} |d_n|^2 < \infty,\end{aligned}$$

where the last equality follows from (48). Defining

$$\nu(\omega) = e^{-i\frac{\omega}{2}}\lambda\left(\frac{\omega}{2}\right)$$

gives the representation (51).

Conversely, suppose that

$$\hat{f}(\omega) = e^{i\omega} m_0(\omega + \pi)^* \nu(2\omega)\widehat{\varphi}(\omega),$$

for some 2π-periodic function ν in $L^2_{\mathbb{C}}([-\pi, +\pi])$. That is,

$$\hat{f}(\omega) = d(\omega)\widehat{\varphi}(\omega),$$

where

$$d(\omega) = e^{i\omega} m_0(\omega + \pi)^*.$$

Since $|m_0(\omega)| \le 1$, this implies that $d(\omega) \in L^2_{\mathbb{C}}([-\pi, +\pi])$. Therefore, $f \in V_0$ (Lemma D3.2). Also, from the expression of $d(\omega)$, $\mathbf{d}_0(\omega) = e^{i\omega}\nu(\omega)\mathbf{m}'_0(\omega)$, and therefore

$$\mathbf{d}_0 \perp \mathbf{m}_0(\omega),$$

that is, $d(\omega)m_0(\omega)^* + d(\omega + \pi)m_0(\omega + \pi)^* = 0$. By Lemma D3.1 and Eq. (48), this implies that $f \perp V_{-1}$. But also $f \in V_0$. Therefore, $f \in W_0$. ■

We are now ready for the main result of this subsection, the Fourier characterization of the mother wavelet in terms of the scaling function and of the high-pass filter.

THEOREM **D3.3.** *The function ψ is a mother wavelet if and only if*

$$\widehat{\psi}(\omega) = e^{i\omega/2} m_0\left(\frac{\omega}{2} + \pi\right)^* \nu(\omega)\widehat{\varphi}\left(\frac{\omega}{2}\right), \tag{53}$$

for some 2π-periodic function ν such that $|\nu(\omega)| = 1$ almost everywhere.

Proof: Since $\widehat{\psi}$ is of the form (50) with $|\nu| = 1$, it is in $L^2_{\mathbb{C}}(\mathbb{R})$ (by the now standard argument) and, therefore, it is the FT of a function $\psi \in L^2_{\mathbb{C}}(\mathbb{R})$, which is in W_0 by Lemma D3.2. By Lemma D3.2 again, any function $g \in W_0$ has an FT of the form

$$\hat{g}(\omega) = e^{i\omega/2} m_0\left(\frac{\omega}{2} + \pi\right)^* s(\omega)\widehat{\varphi}\left(\frac{\omega}{2}\right),$$

for some 2π-periodic function s in $L^2_{\mathbb{C}}([-\pi, +\pi]$. In particular, since $\nu^{-1} = \nu^*$,

$$\hat{g}(\omega) = s(\omega)\nu(\omega)^*\widehat{\psi}(\omega).$$

Since $s\nu^* \in L^2_{\mathbb{C}}([-\pi, +\pi]$,

$$s(\omega)\nu(\omega)^* = \sum_{n\in\mathbb{Z}} c_k e^{-in\omega},$$

for some sequence $\{c_n\}_{n\in\mathbb{Z}} \in \ell^2_{\mathbb{C}}$, and therefore,

$$g(t) = \sum_{n\in\mathbb{Z}} c_k \psi(t-n).$$

Therefore, the translates of ψ generate W_0. The system $\{\psi(\cdot - n)\}_{n\in\mathbb{Z}}$ is orthonormal because

$$\sum_{k\in\mathbb{Z}} |\widehat{\psi}(\omega + 2k\pi)|^2 = 1, \quad \text{a.e.},$$

as can be checked by the usual routine.

Conversely, let ψ be an orthonormal wavelet. Being in W_0, it is of the form (50). By the usual calculations, we find that

$$\sum_{k\in\mathbb{Z}} |\widehat{\psi}(\omega + 2k\pi)|^2 = |\nu(\omega)|^2,$$

and therefore, by the orthonormality condition, $|\nu(\omega)|^2 = 1$. ■

In summary, the mother wavelet is of the form

$$\widehat{\psi}(\omega) = m_1\left(\frac{\omega}{2}\right)\widehat{\varphi}\left(\frac{\omega}{2}\right), \tag{54}$$

where $m_1(\omega) = e^{i\omega} m_0(\omega+\pi)^* \nu(2\omega)$ is a *high-pass* filter ($|m_1(\pi)| = 1$). We recall that the scaling function φ and the low-pass filter are related by

$$\widehat{\varphi}(\omega) = m_0\left(\frac{\omega}{2}\right)\widehat{\varphi}\left(\frac{\omega}{2}\right). \tag{55}$$

We also have the identity

$$|m_0(\omega)|^2 + |m_1(\omega)|^2 = 1. \tag{56}$$

These three relations fully describe the MRA; (55) is called the *dilation equation* and tells us that V_{j-1} is obtained by low-pass filtering V_j; (54) tells us that W_{j-1} is obtained by high-pass filtering V_j; and (56) guarantees that there is no loss of energy.

The choice $\nu \equiv 1$ leads to

$$\widehat{\psi}(\omega) = e^{i\omega/2} m_0\left(\frac{\omega}{2} + \pi\right)^* \widehat{\varphi}\left(\frac{\omega}{2}\right),$$

or, equivalently, up to the sign,

$$\psi(t) = \sqrt{2} \sum_{n\in\mathbb{Z}} (-1)^{n-1} h^*_{-n-1} \varphi(2t - n), \tag{57}$$

where h_n is defined by (40).

EXAMPLE **D3.1** (The Haar Wavelet). *We shall obtain the Haar wavelet by the general method just described. Recall that the scaling function is* $\varphi(t) = 1_{[0,1]}(t)$

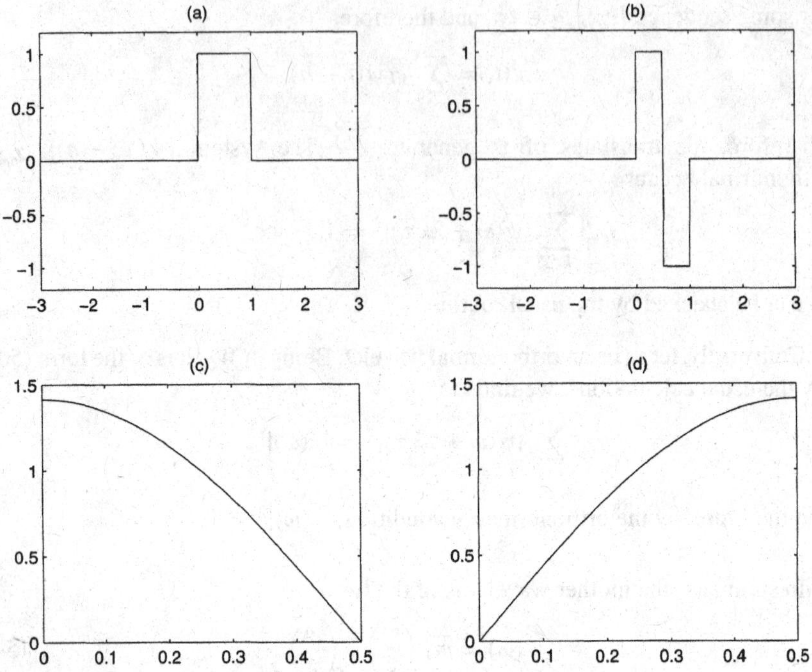

Figure D3.6. Haar scaling function and the corresponding wavelet (left: scaling function; right: wavelet; top: time domain; bottom: frequency domain)

and, therefore,

$$h_n = \sqrt{2} \int_{\mathbb{R}} \varphi(x)\varphi(2x - n)^* \, \mathrm{d}x$$

$$= \begin{cases} \dfrac{1}{\sqrt{2}} & \textit{for } n = 0, 1, \\ 0 & \textit{otherwise,} \end{cases}$$

and, using (57),

$$\psi(t) = \varphi(2t) - \varphi(2t - 1).$$

Thus, we recover the Haar wavelet (Fig. D3.6).

EXAMPLE **D3.2** (Shannon wavelet). *Here*

$$\widehat{\varphi}(\omega) = 1_{[-\pi,+\pi]}(\omega),$$

and, therefore,

$$\varphi(t) = \frac{\sin(\pi t)}{\pi t}.$$

We first choose m_0 *such that* (41) *holds, i.e.,*

$$\widehat{\varphi}(2\omega) = m_0(\omega)\widehat{\varphi}(\omega).$$

Therefore, necessarily,

$$m_0(\omega) = \widehat{\varphi}(2\omega) \quad \text{on} \quad [-\pi\,,\,+\pi],$$

that is,

$$m_0(\omega) = 1_{[-\frac{\pi}{2},+\frac{\pi}{2}]}(\omega) \quad \text{on} \quad [-\pi\,,\,+\pi].$$

By periodicity,

$$m_0(\omega) = \sum_{k\in\mathbb{Z}} \widehat{\varphi}(2\omega + 2k\pi).$$

Our choice of ψ *is as in* (53), *with* $v(\omega) = -ie^{-i\omega}$*:*

$$\begin{aligned}\widehat{\psi}(2\omega) &= -\,e^{-i\omega} m_0(\omega+\pi)^* \widehat{\varphi}(\omega)\\ &= -\,e^{-i\omega}\left(\sum_{k\in\mathbb{Z}} \widehat{\varphi}(2\omega + 2k\pi + 1)\right)\widehat{\varphi}(\omega)\\ &= -\,e^{-i\omega}(\widehat{\varphi}(2\omega+\pi) + \widehat{\varphi}(2\omega-\pi))\\ &= -\,e^{-i\omega}\left(1_{[-\pi,-\frac{\pi}{2}]}(\omega) + 1_{[+\frac{\pi}{2},+\pi]}(\omega)\right).\end{aligned}$$

This gives the Shannon wavelet (Fig. D3.7).

$$\psi(t) = \cos(\tfrac{3}{2}\pi(t-\tfrac{1}{2}))\,\frac{\sin(\frac{1}{2}\pi(t-\frac{1}{2}))}{\frac{1}{2}\pi(t-\frac{1}{2})}.$$

D3·3 Mallat's Algorithm

Mallat's algorithm[6] is a fast algorithm for obtaining from the projection at a given level the wavelet's coefficients at coarser levels of resolution.

Let f be a function in $L^2_{\mathbb{C}}(\mathbb{R})$. Its projection on V_j, the resolution space at level j, is

$$P_j f = \sum_{n\in\mathbb{Z}} c_{j,n}\varphi_{j,n},$$

where

$$c_{j,n} = \langle f, \varphi_{j,n}\rangle. \tag{58}$$

[6]Mallat, S. A theory of multiresolution signal decomposition: The wavelet representation, *IEEE Transactions on Pattern Analysis and Machine Intelligence*, 11, 1989, 674–693.

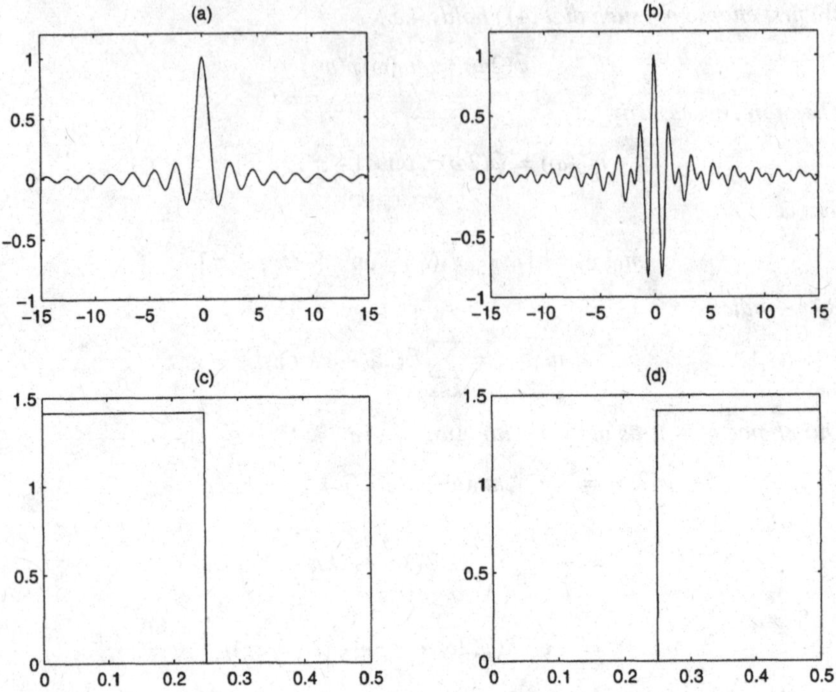

Figure D3.7 Shannon scaling function and the corresponding wavelet (left: scaling function; right: wavelet; top: time domain; bottom: frequency domain)

Its projection on W_j, the space of details at level j, is

$$D_j f = \sum_{n \in \mathbb{Z}} d_{j,n} \psi_{j,n},$$

where

$$d_{j,n} = \langle f, \psi_{j,n} \rangle, \tag{59}$$

and we have

$$P_j f = P_{j-1} f + D_{j-1} f. \tag{60}$$

Denote by $\mathbf{c}_j$ and $\mathbf{d}_j$ the sequences $\{c_{j,n}\}_{n \in \mathbb{Z}}$ and $\{d_{j,n}\}_{n \in \mathbb{Z}}$, respectively. The purpose of Mallat's algorithms is to *decompose* the function f, that is, to pass from $\mathbf{c}_M$ to $\mathbf{d}_{M-1}$, $\mathbf{d}_{M-1}, \ldots, \mathbf{d}_0$, $\mathbf{c}_0$, and to *reconstruct* that is to pass, from $\mathbf{c}_0$, $\mathbf{d}_0$,$\mathbf{d}_1$, ..., $\mathbf{d}_M$ to $\mathbf{c}_M$.

The sequence $\mathbf{d}_{M-1}, \mathbf{d}_{M-1}, \ldots, \mathbf{d}_0, \mathbf{c}_0$ is the *wavelet encoding* of the *wavelet data* $\mathbf{c}_M$. We shall explain the interest of this encoding once we have derived Mallat's algorithm.

Since the function $\varphi(t/2)$ is in V_{-1}, and $V_{-1} \subset V_0$, and since $\{\varphi(\cdot - n)\}_{n\in\mathbb{Z}}$ is a Hilbert basis of V_0, we have the decomposition

$$\frac{1}{2}\varphi(\frac{1}{2}t) = \sum_{n\in\mathbb{Z}} \alpha_n \varphi(t+n),$$

where

$$\alpha_n = \frac{1}{2}\int_{\mathbb{R}} \varphi(\frac{1}{2}t)\varphi(t+n)\,dt.$$

Therefore,

$$\begin{aligned}\varphi_{j-1,n}(t) &= 2^{\frac{j-1}{2}}\varphi(2^{j-1}t - n)\\ &= 2^{\frac{j-1}{2}}\frac{1}{2}\varphi(\frac{1}{2}(2^j t - 2n))\\ &= 2^{\frac{j-1}{2}}\sum_{k\in\mathbb{Z}} \alpha_k \varphi(2^j t - 2n + k),\end{aligned}$$

that is,

$$\varphi_{j-1,n} = \sqrt{2}\sum_{k\in\mathbb{Z}} \alpha_k \varphi_{j,2n-k}. \tag{61}$$

Similarly, since the function $\psi(t/2)$ is in W_{-1}, and $W_{-1} \subset V_0$, and since $\{\varphi(\cdot - n)\}_{n\in\mathbb{Z}}$ is a Hilbert basis of V_0, we have the decomposition

$$\frac{1}{2}\psi(\frac{1}{2}t) = \sum_{n\in\mathbb{Z}} \beta_n \varphi(t+n),$$

where

$$\beta_n = \frac{1}{2}\int_{\mathbb{R}} \psi(\frac{1}{2}t)\varphi(t+n)\,dt.$$

Therefore, it follows by computations similar to those above that

$$\psi_{j-1,n} = \sqrt{2}\sum_{k\in\mathbb{Z}} \beta_k \varphi_{j,2n-k}. \tag{62}$$

Denoting the low-pass filter by

$$m_0(\omega) = \sum_{n\in\mathbb{Z}} \alpha_n e^{in\omega}$$

and the high-pass filter by

$$m_1(\omega) = \sum_{n\in\mathbb{Z}} \beta_n e^{in\omega},$$

we have, from (55) that

$$\widehat{\varphi}(2\omega) = m_0(\omega)\varphi(\omega),$$

and, from (54) that

$$\widehat{\psi}(2\omega) = m_1(\omega)\varphi(\omega).$$

In Theorem D3.3, we now make the particular choice of the mother wavelet corresponding to $\nu(\omega) = 1$:

$$m_1(\omega) = e^{i\omega} m_0(\omega + \pi)^*,$$

that is,

$$\sum_{n\in\mathbb{Z}} \beta_n e^{in\omega} = \sum_{n\in\mathbb{Z}} (-1)^{n+1} \alpha_{1-n}^* e^{in\omega}.$$

Therefore,

$$\beta_n = (-1)^{n+1} \alpha_{1-n}^*. \tag{63}$$

Substituting (61) in (58), we obtain

$$c_{j-1,n} = \sqrt{2} \sum_{k\in\mathbb{Z}} \alpha_k^* c_{j,2n-k}. \tag{64}$$

Similarly, substituting (62) in (59), we obtain

$$d_{j-1,n} = \sqrt{2} \sum_{k\in\mathbb{Z}} \beta_k^* c_{j,2n-k}. \tag{65}$$

These are the basic recursions of the *decomposition algorithm* (see Fig. D3.8). The recursions for the *reconstruction algorithm* (see Fig. D3.9) are obtained from (60), (61), and (62). This gives

$$c_{j,n} = \sqrt{2} \sum_{k\in\mathbb{Z}} \left[\alpha_{2k-n} c_{j-1,k} + \beta_{2k-n} d_{j-1,k}\right]. \tag{66}$$

EXERCISE **D3.1.** *Show that for the Haar wavelet,*

$$c_{j-1,k} = \frac{c_{j,2k-1} + c_{j,2k}}{\sqrt{2}}.$$

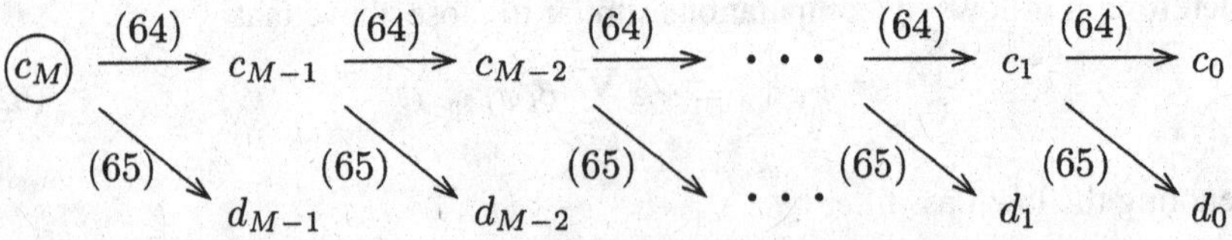

Figure D3.8. Mallat's decomposition algorithm

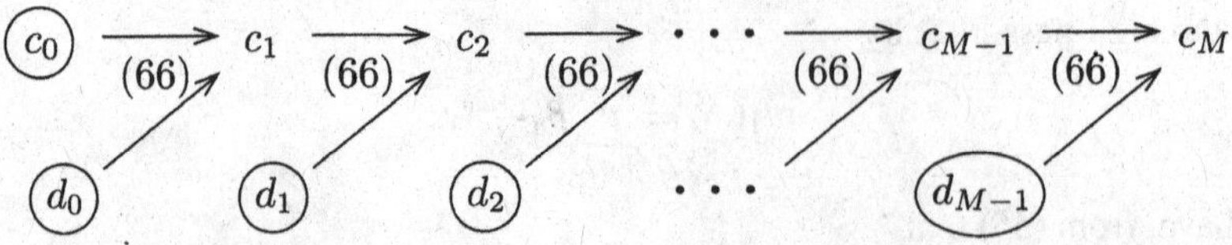

Figure D3.9. Mallat's reconstruction algorithm

and

$$d_{j-1,k} = \frac{c_{j,2k-1} - c_{j,2k}}{\sqrt{2}}.$$

We shall now evaluate the algorithmic complexity of the decomposition algorithm. (Similar results hold for the reconstruction algorithm.)

For this we suppose that the low-pass and high-pass filters, m_0 and m_1, respectively, have *finite* impulse responses, that is, the sequences $\{\alpha_n\}_{n\in\mathbb{Z}}$ and $\{\beta_n\}_{n\in\mathbb{Z}}$ have a finite-number (say, K) of nonzero terms. Suppose that the infinite-dimensional vector $\mathbf{c}_M$ has in practice a finite number N of nonzero terms (say, after truncation). Then there are approximately $N/2$ terms in $\mathbf{c}_{M-1}$, and therefore, in view of (64), the passage from $\mathbf{c}_M$ to $\mathbf{c}_{M-1}$ costs approximately $KN/2$ multiplications; so does the passage from $\mathbf{c}_M$ to $\mathbf{d}_{M-1}$. For the decomposition algorithm, we therefore have approximately

$$\left(K\frac{N}{2} + K\frac{N}{4} + \ldots + K\frac{N}{2^M}\right) + K\frac{N}{2^M} = KN$$

multiplications. The complexity of Mallat's algorithm is therefore linear in data size.

Note that Mallat's algorithm encodes N numbers into N numbers. Thus the compression gain seems to be null. However, only a few terms in the sequence of details $d_{j,\ell}$, $\ell \in \mathbb{Z}$, $j = M-1, \ldots, 0$, are nonnegligible, provided the MRA is sufficiently smooth. The smoothness issue is discussed in Chapter D5.

D4

Construction of an MRA

D4·1 MRA from an Orthonormal System

The Fourier structure of an MRA is now elucidated, and we know how to obtain a wavelet basis when an MRA is given. This chapter gives two methods for obtaining an MRA.

In the previous chapter, we started from a nested family of resolution spaces $\{V_j\}_{j\in\mathbb{Z}}$ and we discovered a scaling function φ in rather simple examples. Now, obtaining the scaling function from a *given* nested family of resolutions spaces can be a difficult task in general. However, if we are interested in a wavelet basis rather than in a given family of resolution spaces, we might as well start from a *given* function $\varphi \in L^2$ with the property that $\{\varphi(\cdot\ - n)\}_{n\in\mathbb{Z}}$ is an orthonormal system, and define the resolution spaces in an ad hoc manner guaranteeing that φ is indeed the corresponding scaling function.

If φ is to be the scaling function, there is but one choice for the resolution spaces, namely,

$$V_j = \overline{\text{span}}\,\{\varphi_{j,n} : n \in \mathbb{Z}\}.$$

An inspired choice of φ will make the V_j's nested as required, and this has to be verified because there is no reason why it should be so when one starts from an arbitrary orthonormal system $\{\varphi(\cdot\ - n)\}_{n\in\mathbb{Z}}$. A necessary and sufficient condition for this is that

$$\varphi(t) = \sum_{n\in\mathbb{Z}} c_n \varphi(2t - n), \tag{67}$$

for some sequence $\{c_n\}_{n\in\mathbb{Z}} \in \ell^2_{\mathbb{C}}(\mathbb{Z})$ or, equivalently, that the dilation equation

$$\widehat{\varphi}(\omega) = m_0\left(\frac{\omega}{2}\right)\widehat{\varphi}\left(\frac{\omega}{2}\right) \tag{68}$$

holds for some 2π-periodic function m_0 in $L^2_{\mathbb{C}}(-\pi, +\pi)$. We must also verify that conditions (d) in the definition of, an MRA are satisfied. By Theorem D3.1, it suffices that $\widehat{\varphi}$ be continuous at the origin and that $|\widehat{\varphi}(0)| = 1$.

Meyer's Wavelet

Define φ by

$$\widehat{\varphi}(\omega) = \begin{cases} 1 & \text{if } |\omega| \leq \dfrac{2\pi}{3}, \\ \cos\left(\dfrac{\pi}{2}\,\nu\left(\dfrac{3}{4\pi}\,|\omega| - 1\right)\right) & \text{if } \dfrac{2\pi}{3} \leq |\omega| \leq \dfrac{4\pi}{3}, \\ 0 & \text{otherwise,} \end{cases} \tag{69}$$

where ν is a smooth function (C^k or C^∞) such that

$$\nu(x) = \begin{cases} 0 & \text{if } x \leq 0, \\ 1 & \text{if } x \geq 1 \end{cases} \tag{70}$$

and

$$\nu(x) + \nu(1-x) = 1. \tag{71}$$

Using (71) it is easy to verify that

$$\sum_{k\in\mathbb{Z}} |\widehat{\varphi}(\omega + 2k\pi)|^2 = 1,$$

and, therefore, $\{\varphi(\cdot - n)\}_{n\in\mathbb{Z}}$ is an orthonormal system. We must now verify that the V_j are nested, and for this it suffices to verify that $V_0 \subset V_1$ or, equivalently, that $\varphi \in V_1$. But this is true if and only if there exists a 2π-periodic function m_0 of finite power such that

$$\widehat{\varphi}(\omega) = m_0\left(\frac{\omega}{2}\right)\widehat{\varphi}\left(\frac{\omega}{2}\right).$$

It turns out that

$$m_0(\omega) = \sum_{k\in\mathbb{Z}} \widehat{\varphi}(2\omega + 4k\pi)$$

accomplishes what is required. In fact,

$$\begin{aligned} m_0\left(\frac{\omega}{2}\right)\widehat{\varphi}\left(\frac{\omega}{2}\right) &= \sum_{k\in\mathbb{Z}} \widehat{\varphi}(\omega + 2k\pi)\widehat{\varphi}\left(\frac{\omega}{2}\right) \\ &= \widehat{\varphi}(\omega)\widehat{\varphi}\left(\frac{\omega}{2}\right), \end{aligned}$$

since the supports of $\widehat{\varphi}(\omega+2k\pi)$ and of $\widehat{\varphi}(\omega/2)$ do not overlap if $k \neq 0$. But since

$$\widehat{\varphi}\left(\frac{\omega}{2}\right) = 1 \quad \text{if } \omega \in \operatorname{supp}(\widehat{\varphi}),$$

we have

$$\widehat{\varphi}(\omega)\widehat{\varphi}\left(\frac{\omega}{2}\right) = \widehat{\varphi}(\omega),$$

as desired. We obtain a mother wavelet by formula (53) of with $\nu(\omega) = 1$. This gives

$$\begin{aligned}\widehat{\psi}(\omega) &= e^{i\omega/2} m_0\left(\frac{\omega}{2}+\pi\right)^* \widehat{\varphi}\left(\frac{\omega}{2}\right)\\ &= e^{i\omega/2}\sum_{k\in\mathbb{Z}} \widehat{\varphi}(\omega+2\pi+4k\pi)\widehat{\varphi}\left(\frac{\omega}{2}\right)\\ &= e^{i\omega/2}(\widehat{\varphi}(\omega+2\pi)+\widehat{\varphi}(\omega-2\pi))\widehat{\varphi}\left(\frac{\omega}{2}\right),\end{aligned}$$

which gives

$$\widehat{\psi}(\omega) = \begin{cases} e^{i\omega/2}\sin\left(\frac{\pi}{2}\nu\left(\frac{3}{2\pi}|\omega|-1\right)\right) & \text{if } \frac{2\pi}{3} \le |\omega| \le \frac{4\pi}{3},\\ e^{i\omega/2}\cos\left(\frac{\pi}{2}\nu\left(\frac{3}{2\pi}|\omega|-1\right)\right) & \text{if } \frac{4\pi}{3} \le |\omega| \le \frac{8\pi}{3},\\ 0 & \text{otherwise.}\end{cases} \tag{72}$$

EXERCISE **D4.1.** *Let P be a probability measure on $\mathbb{R}$ with support in $[-\varepsilon,+\varepsilon] \subset [-\frac{\pi}{3},+\frac{\pi}{3}]$, and define $\varphi(t)$ by its Fourier transform*

$$\widehat{\varphi}(\omega) = \left(\int_{\omega-\pi}^{\omega+\pi} \mathrm{d}P\right)^{\frac{1}{2}}.$$

Check that $\varphi(t)$ is indeed in $L^2_{\mathbb{C}}(\mathbb{R})$ and that the system $\{\varphi(\cdot - n]\}_{n\in\mathbb{Z}}$ is orthonormal. Check that the dilation equation (55) holds with

$$m_0\left(\frac{\omega}{2}\right) = \begin{cases} \widehat{\varphi}(\omega) & \text{if } |\omega| \le \frac{4\pi}{3},\\ 0 & \text{otherwise.}\end{cases}$$

Show that $\varphi(t)$ so defined is the scaling function of some multiresolution analysis and that a mother wavelet is given by its Fourier transform

$$\widehat{\psi}(\omega) = e^{-i\frac{\omega}{2}}\left(\int_{|\frac{\omega}{2}|-\pi}^{|\omega|-\pi} \mathrm{d}P\right)^{\frac{1}{2}}.$$

Examine the case where P is the Dirac measure at 0.

D4·2 MRA from a Riesz Basis

Now we do not impose orthonormality. To be specific, we have an L^2-function w such that

$$w(t) = \sum_{n\in\mathbb{Z}} c_n w(2t - n), \tag{73}$$

where $\{c_n\}_{n\in\mathbb{Z}} \in \ell^2_{\mathbb{C}}(\mathbb{Z})$, and we *define* the resolution spaces by

$$V_j = \overline{\operatorname{span}}\,\{w_{j,n} : n \in \mathbb{Z}\}. \tag{74}$$

Of course, condition (73) guarantees that these spaces are nested. In order to obtain a Hilbert basis of V_0, we use Theorem C4.6 which says that under the "frame" condition

$$0 < \alpha \le \sum_{k\in\mathbb{Z}} |\widehat{w}(\omega + 2k\pi)|^2 \le \beta < \infty, \tag{75}$$

the system $\{\varphi(\cdot - n)\}_{n\in\mathbb{Z}}$ is a Hilbert basis of V_0, where

$$\widehat{\varphi}(\omega) = \frac{\widehat{w}(\omega)}{\sum_{k\in\mathbb{Z}} |\widehat{w}(\omega + 2k\pi)|^2}. \tag{76}$$

Here we shall also have to verify that

$$\bigcap_{j\in\mathbb{Z}} V_j = \varnothing \quad \text{and} \quad \operatorname{clos}\left(\bigcup_{j\in\mathbb{Z}} V_j\right) = L^2.$$

For this we can use the following result.

THEOREM **D4.1.** *Let $w \in L^2_{\mathbb{C}}(\mathbb{R})$ satisfy*

$$0 < \alpha \le \sum_{k\in\mathbb{Z}} |\widehat{w}(\omega + 2k\pi)|^2 \le \beta < \infty, \tag{77}$$

and define

$$V_j = \overline{\operatorname{span}}\,\{w_{j,k} : k \in \mathbb{Z}\}. \tag{78}$$

Suppose that the V_j are nested. Then

$$\bigcap_{j\in\mathbb{Z}} V_j = \varnothing. \tag{79}$$

Proof: The inequalities (77) are equivalent to the existence of $A > 0$, $B < \infty$ such that

$$0 < A\|f\|^2 \le \sum_{k\in\mathbb{Z}} |\langle f, w_{0,k}\rangle|^2 \le B\|f\|^2, \tag{80}$$

for all $f \in V_0$, and therefore equivalent to

$$0 < A\|f\|^2 \le \sum_{k\in\mathbb{Z}} |\langle f, w_{j,k}\rangle|^2 \le B\|f\|^2 < \infty,$$

for all $f \in V_j$, $f \neq 0$, and all $j \in \mathbb{Z}$.

With any $f \in \bigcap_{j\in\mathbb{Z}} V_j$ and $\varepsilon > 0$, one can associate a compactly supported and continuous function $\tilde{f} \in L^2_{\mathbb{C}}(\mathbb{R})$ such that $\|f - \tilde{f}\| \leq \varepsilon$, and therefore on denoting the orthogonal projection on V_j by P_j, we have

$$\|f - P_j\tilde{f}\| = \|P_j(f - \tilde{f})\| \leq \|f - \tilde{f}\| \leq \varepsilon.$$

Therefore, for all $j \in \mathbb{Z}$,

$$\|f\| \leq \varepsilon + \|P_j\tilde{f}\|.$$

By (80),

$$\begin{aligned}\|P_j\tilde{f}\| &\leq A^{-1/2}\left[\sum_{k\in\mathbb{Z}} |\langle P_j\tilde{f}, w_{jk}\rangle|^2\right]^{1/2} \\ &= A^{-1/2}\left[\sum_{k\in\mathbb{Z}} |\langle \tilde{f}, w_{jk}\rangle|^2\right]^{1/2}.\end{aligned}$$

Now, with $c > 0$ such that

$$\operatorname{supp} \tilde{f} \subset [-c, +c] \quad \text{and} \quad \widetilde{M} = \sup_{x\in\mathbb{R}} |\tilde{f}(x)|,$$

we have

$$\begin{aligned}|\langle \tilde{f}, w_{jk}\rangle|^2 &= \left|2^{-j}\int_{|x|<c} |\tilde{f}(x)w(2^{-j}x - k)|\,\mathrm{d}x\right|^2 \\ &\leq 2^{-j}\left(\int_{|x|<c} |\tilde{f}(x)w(2^{-j}x - k)|\,\mathrm{d}z\right)^2 \\ &\leq 2^{-j}\widetilde{M}^2\left(\int_{|x|<c} |w(2^{-j}x - k)|\,\mathrm{d}x\right)^2 \\ &\leq 2^{-j}\widetilde{M}^2 2c\int_{|x|<c} |w(2^{-j}x - k)|^2\,\mathrm{d}x,\end{aligned}$$

where the last inequality is Schwarz's inequality. Therefore,

$$|\langle \tilde{f}, w_{jk}\rangle|^2 \leq 2c\widetilde{M}^2\int_{k-2^{-j}c}^{k+2^{-j}c} |w(x)|^2\,\mathrm{d}x.$$

Assuming j to be large enough for $2^{-j} \leq 1/2$ to hold, and then summing with respect to $k \in \mathbb{Z}$,

$$\sum_{k\in\mathbb{Z}} |\langle \tilde{f}, w_{jk}\rangle|^2 \leq 2c\widetilde{M}^2\int_{A(c,j)} |w(x)|^2\,\mathrm{d}x,$$

where

$$A(c, j) = \sum_{k\in\mathbb{Z}} [k - 2^{-j}c, k + 2^{-j}c].$$

By the dominated convergence theorem this term tends to 0 as $j \to \infty$. In particular, there exists a j such that $\|P_j \tilde{f}\| \le \varepsilon$, and therefore $\|f\| \le 2\varepsilon$. Since ε is arbitrary, this implies $\|\tilde{f}\| = 0$, which proves (79). ■

Let us now see how this program works in a classic example:

Franklin's Wavelet

Take[7] for w the *piecewise linear spline*

$$w(x) = \begin{cases} 1 - |x| & \text{if } 0 \le |x| \le 1, \\ 0 & \text{otherwise}, \end{cases} \tag{81}$$

and observe that (73) is verified. More explicitly,

$$w(x) = \tfrac{1}{2} w(2x+1) + w(2x) + \tfrac{1}{2} w(2x-1). \tag{82}$$

The Fourier transform of w is

$$\widehat{w}(\omega) = \left(\frac{\sin(\frac{\omega}{2})}{\frac{\omega}{2}} \right)^2 .$$

We have

$$\sum_{k \in \mathbb{Z}} |\widehat{w}(\omega + 2k\pi)|^2 = \tfrac{2}{3} + \tfrac{1}{3}\cos(\omega)$$

$$= \tfrac{1}{3}\left(1 + 2\cos^2\left(\frac{\omega}{2}\right)\right).$$

(One way to prove this is to compute the Fourier coefficients of the left-hand side

$$\frac{1}{2\pi} \int_0^{2\pi} \sum_{k \in \mathbb{Z}} |\widehat{w}(\omega + 2k\pi)|^2 e^{-in\omega}\, d\omega = \frac{1}{2\pi} \int_{\mathbb{R}} |\widehat{w}(\omega)|^2 e^{-in\omega}\, d\omega$$

$$= \int_{\mathbb{R}} w(t) w(t+n)^*\, dt,$$

and this immediately gives the result. Note the generality of the method and its interest when $w(t)$ is compactly supported.) The mother wavelet is then obtained from (76). This gives

$$\widehat{\varphi}(\omega) = \widehat{w}(\omega) \frac{\sqrt{3}}{\left(1 + 2\cos^2\left(\frac{\omega}{2}\right)\right)^{1/2}} . \tag{83}$$

If we can compute, at least numerically, the Fourier coefficient c_n in

$$\sqrt{3}\left(1 + 2\cos^2\left(\frac{\omega}{2}\right)\right)^{-1/2} = \sum_{n \in \mathbb{Z}} c_n e^{-in\omega},$$

---

[7] Franklin, P. A set of continuous orthogonal functions, *Math. Ann.*, 100, 1928, 522–529.

then we obtain $\varphi(t)$ as

$$\varphi(t) = \sum_{n\in\mathbb{Z}} c_n w(t-n).$$

The corresponding low-pass filter $m_0(\omega)$ is

$$m_0(\omega) = \frac{\widehat{\varphi}(2\omega)}{\widehat{\varphi}(\omega)} = \cos^2\left(\frac{\omega}{2}\right)\left(\frac{1+2\cos^2\left(\frac{\omega}{2}\right)}{1+2\cos^2(\omega)}\right)^{1/2},$$

and this leads to an expression for the mother wavelet's Fourier transform. Again the (numerical) evaluation of the Fourier coefficients of the function factoring $\widehat{\varphi}(\omega)$ yields an evaluation of $\psi(t)$ in terms of the translates of $\varphi(2x)$.

D4·3 Spline Wavelets

Franklin's wavelet is a particular case of the Battle–Lemarié spline wavelets, which are now described. We first introduce a family of functions, the *basis splines*, or *B-splines*, that play an important role in numerical analysis, in the theory of spline approximation. The B-spline functions $B_n(t)$, $n \geq 0$, are defined recursively by

$$B_0(t) = 1_{[0,1]}(t)$$

and, for $n \geq 1$,

$$B_{n+1}(t) = (B_0 * B_n)(t) = \int_{t-1}^{t} B_n(x)\,\mathrm{d}x.$$

For $n = 3$, we have

$$B_3(t) = \begin{cases} \frac{1}{6}t^3 & \text{if } 0 \leq t \leq 1, \\ \frac{2}{3} - 2t + 2t^2 - \frac{1}{2}t^3 & \text{if } 1 \leq t \leq 2, \\ 0 & \text{if } t < 0, \end{cases}$$

the rest of the function being obtained by symmetry around 2.

In the general case, $B_n(t)$ is (for $n \geq 1$) in C^{n-1}, its support is the interval $[0, n+1]$, and

$$\int_{\mathbb{R}} B_n(x)\,\mathrm{d}x = 1.$$

We have

$$\hat{B}_0(\omega) = e^{-i\frac{\omega}{2}}\operatorname{sinc}\left(\frac{\omega}{2\pi}\right),$$

and, therefore, in the Fourier domain, the recurrence defining the B-splines becomes by the convolution–multiplication rule

$$\hat{B}_{n+1}(\omega) = e^{-i\frac{\omega}{2}}\operatorname{sinc}\left(\frac{\omega}{2\pi}\right)\hat{B}_n(\omega).$$

This gives, for $n \geq 0$,

$$\hat{B}_n(\omega) = \left(e^{-i\frac{\omega}{2}} \operatorname{sinc}\left(\frac{\omega}{2\pi}\right)\right)^{n+1}. \tag{84}$$

We observe that $\hat{B}_n(0) = 1$ and that in the neighborhood of 0

$$\hat{B}_n(\omega) = O\left(\frac{1}{|\omega|^{n+1}}\right). \tag{85}$$

We shall now seek a scaling function for the B-spline of order n. From the observation

$$e^{-i\frac{\omega}{2}} \operatorname{sinc}\left(\frac{\omega}{2\pi}\right) = e^{-i\frac{\omega}{4}} \cos\left(\frac{\omega}{4\pi}\right) \times e^{-i\frac{\omega}{4}} \operatorname{sinc}\left(\frac{\omega}{4\pi}\right),$$

it follows that

$$\hat{B}_n(\omega) = m_0\left(\frac{\omega}{2}\right) \hat{B}_n\left(\frac{\omega}{2}\right),$$

where

$$m_0(\omega) = \left(e^{-i\frac{\omega}{2}} \cos\left(\frac{\omega}{2}\right)\right)^{n+1} = \left(\frac{1+e^{-i\omega}}{2}\right)^{n+1}.$$

The impulse response of the low-pass filter of the MRA is therefore

$$h_k = \frac{\sqrt{2}}{2^{n+1}}\binom{n+1}{k}, \qquad 0 \leq k \leq n+1,$$

and the scaling equation is

$$B_n(t) = \frac{1}{2^n} \sum_{k=0}^{n+1} \binom{n+1}{k} B_n(2t-k).$$

In view of the estimate (85), the series

$$\sum_{k \in \mathbb{Z}} |\hat{B}_n(\omega + 2k\pi)|^2 \tag{86}$$

is absolutely convergent. Using (84) and the estimate

$$\frac{\sin(\omega)}{\omega} \geq \frac{2}{\pi}, \qquad 0 \leq \omega \leq \frac{\pi}{2},$$

we have, for $|\omega| \leq \pi$,

$$\left|\hat{B}_n\left(\frac{\omega}{2}\right)\right|^2 = \left|\frac{\sin(\frac{\omega}{2})}{\frac{\omega}{2}}\right|^{2n+2} \geq \left(\frac{2}{\pi}\right)^{2n+2}.$$

Therefore, there exist positive finite constants A and B such that

$$A \leq \sum_{k \in \mathbb{Z}} |\hat{B}_n(\omega + 2k\pi)|^2 \leq B,$$

and $\{B_n(\cdot - k)\}_{k \in \mathbb{Z}}$ constitutes a Riesz basis of the Hilbert subspace that it generates. In order to compute the scaling function of the MRA, we need the following lemma.

LEMMA D4.1. *There exists a polynomial P_n of degree n such that*

$$\sum_{k\in\mathbb{Z}} |\hat{B}_n(\omega + 2k\pi)|^2 = P_n(\cos(\omega)). \tag{87}$$

Moreover, the coefficients of this polynomial are rational and can be computed recursively.

Proof: Denote the left-hand side of (87) by $F_n(\omega)$. Inserting (86) in(84) gives

$$F_n(\omega) = \left(\sin\left(\frac{\omega}{2}\right)\right)^{2n+2} G_n(\omega),$$

where

$$G_n(\omega) = \sum_{k\in\mathbb{Z}} \frac{1}{\left(\frac{\omega}{2} + \pi k\right)^{2n+2}}.$$

One verifies easily that, for $n \geq 1$,

$$G_n(\omega) = \frac{2}{n(2n+1)} G_n''(\omega),$$

and, therefore,

$$F_n(\omega) = \frac{2}{n(2n+1)} \left(\sin\left(\frac{\omega}{2}\right)\right)^{2n+2} \left(\frac{F_{n-1}(\omega)}{\left(\sin\left(\frac{\omega}{2}\right)\right)^{2n}}\right)''.$$

We introduce the new variable $y = \cos(\omega)$, and define the function P_n by $F_n(\omega) = P_n(y)$. Since $F_0(\omega) = 1$, we have $P_0(y) = 1$. The recursion in the last display becomes

$$P_n(y) = \frac{2}{n(2n+1)}(1-y)^{n+1} \frac{\mathrm{d}}{\mathrm{d}\omega} \left(\frac{P_{n-1}(y)}{(1-y)^n}\right).$$

Using the differentiation rules

$$\frac{\mathrm{d}}{\mathrm{d}\omega} = (-\sin(\omega))\frac{\mathrm{d}}{\mathrm{d}y}$$

and

$$\frac{\mathrm{d}^2}{\mathrm{d}\omega^2} = (-y)\frac{\mathrm{d}}{\mathrm{d}y} + (1-y^2)\frac{\mathrm{d}^2}{\mathrm{d}y^2},$$

we obtain, after simplification and rearrangement,

$$P_n(y) = \frac{2}{n(2n+1)}(n(n+1+ny)P_{n-1}(y) + (1-y)(2n+(2n-1)y)P_{n-1}'$$
$$+ (1-y)^2(1+y)P_{n-1}'').$$

Therefore, if P_{n-1} is a polynomial of degree $n-1$, then P_n is a polynomial of degree n. The conclusion follows since P_0 is indeed a constant. ■

The general method of the previous section gives for scaling function $\widehat{\varphi} = \widehat{\varphi}_n$

$$\widehat{\varphi}(\omega) = \frac{\hat{B}_n(\omega)}{P_n(\cos(\omega))^{\frac{1}{2}}}.$$

Therefore,

$$\varphi(t) = \sum_{k \in \mathbb{Z}} c_k B_n(t-k),$$

where the c_k's are given by the power-series expansion

$$\frac{1}{\left(P_n\left(\frac{z+z^{-1}}{2}\right)\right)^{\frac{1}{2}}} = \sum_{k \in \mathbb{Z}} c_k z^k.$$

Observe that $c_k = c_{-k}$. Also, since the function in the left-hand side is analytic,

$$|c_k| \leq \rho^{|k|},$$

for some $|\rho| < 1$. In particular, the scaling function $\varphi(t)$ has exponential decay.

We now proceed to compute the mother wavelet. We have to compute the impulse response of the low-pass filter $m_0(\omega)$. We have

$$m_0(\omega) = \frac{\varphi(2\omega)}{\varphi(\omega)} = \frac{\hat{B}_n(2\omega)}{\hat{B}_n(\omega)} \left(\frac{P_n(\cos(\omega))}{P_n(\cos(2\omega))}\right)^{\frac{1}{2}},$$

that is,

$$m_0(\omega) = \left(\frac{1+e^{-in\omega}}{2}\right)^{n+1} \left(\frac{P_n(\cos(\omega))}{P_n(\cos(2\omega))}\right)^{\frac{1}{2}}.$$

We compute the Fourier expansion

$$\left(\frac{P_n(\cos(\omega))}{P_n(\cos(2\omega))}\right)^{\frac{1}{2}} = \sum_{k \in \mathbb{Z}} q_k e^{-ik\omega},$$

where

$$q_k = q_{-k} = \frac{}{\pi} \int_0^{\pi} \left(\frac{P_n(\cos(\omega))}{P_n(\cos(2\omega))}\right)^{\frac{1}{2}} \cos(k\omega)\, d\omega.$$

Therefore,

$$h_r = \frac{\sqrt{2}}{2^{n+1}} \sum_{\ell=0}^{n+1} \binom{n+1}{\ell} q_{r-\ell}.$$

The mother wavelet is then

$$\psi(t) = \sqrt{2} \sum_{k \in \mathbb{Z}} (-1)^{k-1} h_{-k-1} \varphi(2t-k).$$

Putting all this together, we finally obtain

$$\psi(t) = \sum_{k\in\mathbb{Z}} b_k B_n(2t - k),$$

where

$$b_r = \sqrt{2}\sum_{k\in\mathbb{Z}}(-1)^{k-1} h_{-k-1} q_{r-k}.$$

D5

Smooth Multiresolution Analysis

D5·1 Autoreproducing Property of the Resolution Spaces

The axiomatic framework of multiresolution analysis is Fourier analysis in L^2, and the convergence of the wavelet expansion is therefore in the L^2-norm. The smoothness properties of the scaling function and of the mother wavelet are, however, of great interest to obtain fast L^2-convergence of the wavelet expansion, or to obtain pointwise convergence of this expansion.

We first recall a definition.

DEFINITION **D5.1.** *Let $r \geq 0$ be an integer. The function $\varphi : \mathbb{R} \to \mathbb{C}$ is said to be in S_r if for all $n \in \mathbb{N}$, all $0 \leq k \leq r$, there exist finite nonnegative constants $C_{k,n}$ such that, for all $x \in \mathbb{R}$,*

$$|\varphi^{(k)}(x)| \leq \frac{C_{k,n}}{(1+|x|)^n}. \tag{88}$$

DEFINITION **D5.2.** *Consider an MRA with scaling function φ. Let r be an integer. The MRA is called r-*smooth *if $\varphi \in S_r$.*

We now consider an r-smooth multiresolution analysis. We associate with it the function $q : \mathbb{R} \times \mathbb{R} \to \mathbb{C}$, called the *kernel* of the MRA and defined by

$$q(x,t) = \sum_{n\in\mathbb{Z}} \varphi^*(x-n)\varphi(t-n). \tag{89}$$

Using the inequality

$$(1+|a|)(1+|b|) \geq 1+|b-a|,$$

we have

$$\begin{aligned}|q(x,t)| &\leq \sum_{n\in\mathbb{Z}} |\varphi(x-n)||\varphi(t-n)| \\ &\leq \sum_{n\in\mathbb{Z}} \frac{C_{0,k+2}}{(1+|x-n|)^{k+2}} \frac{C_{0,k+2}}{(1+|t-n|)^{k+2}} \\ &\leq C_{0,k+2}^2 \sum_{n\in\mathbb{Z}} \frac{1}{(1+|x-n|)^2} \frac{1}{(1+|t-n|)^2} \frac{1}{(1+|x-t|)^k}.\end{aligned}$$

It follows that, for all $k \in \mathbb{N}$, there exists a finite nonnegative constant C_k such that, for all $x, t \in \mathbb{R}$,

$$|q(x,t)| \leq \frac{C_k}{(1+|x-t|)^k}. \tag{90}$$

In particular, for each $t \in \mathbb{R}$, the function $q_t : \mathbb{R} \to \mathbb{C}$ defined by $q_t(x) = q(x,t)$ is in L^2, and the development of any function $f \in V_0$ along the orthonormal basis $\{\varphi_n\}_{n\in\mathbb{Z}} = \{\varphi(\cdot - n)\}_{n\in\mathbb{Z}}$

$$f = \sum_{n\in\mathbb{Z}} \langle f, \varphi_n \rangle \varphi_n$$

takes the form

$$f(t) = \int_{\mathbb{R}} q(x,t) f(x)\,\mathrm{d}x. \tag{91}$$

DEFINITION **D5.3.** *Let E be some set, and let H be a Hilbert space of functions $f : E \to \mathbb{C}$ with the Hermitian product $\langle\cdot,\cdot\rangle$. If there exists a function $K : E \times E \to \mathbb{C}$ such that for each $x \in E$, the function $K(x,\cdot) \in H$, and $f(x) = \langle K(x,\cdot), f\rangle$, H is called an* autoreproducing Hilbert space *with* reproducing kernel K.

EXERCISE **D5.1.** *Let E be some set, and let H be a Hilbert space of functions $f : E \to \mathbb{C}$ with the Hermitian product $\langle\cdot,\cdot\rangle$. Suppose that for each $x \in E$, the mapping $f \to f(x)$ from H to $\mathbb{C}$ is continuous. Show that H is then an autoreproducing Hilbert space.*

Equation (91) therefore tells us that V_0 is an autoreproducing Hilbert space with reproducing kernel $q(x,t)$. Similarly, for all $m \in \mathbb{Z}$, V_m is an autoreproducing Hilbert space with reproducing kernel $q_m(x,t)$, where

$$q_m(x,t) = 2^m q_m(2^m x, 2^m t). \tag{92}$$

We know (Theorem D3.1) that $|\widehat{\varphi}(0)| = 1$, and we can assume without loss of generality that $\widehat{\varphi}(0) = 1$. Therefore, in view of property (38),

$$\widehat{\varphi}(2k\pi) = 1_{\{k=0\}}. \tag{93}$$

It follows from this and the weak Poisson formula (Theorem A2.3) that

$$\sum_{n\in\mathbb{Z}} \varphi(x-n) = 1. \tag{94}$$

Finally, from (89) and (94),

$$\int_{\mathbb{R}} q_m(x,t)\,\mathrm{d}x = 1, \tag{95}$$

for $m = 1$, and therefore for all $m \in \mathbb{Z}$.

EXERCISE **D5.2.** *Give the kernel $q(x,t)$ of the Haar MRA ($\varphi(t) = 1_{[0,1]}(t)$) and of the Shannon MRA ($\varphi(t) = \sin(\pi t)/\pi t$).*

In general, the kernel of an MRA does not have a closed form, and the examples in the previous exercise are exceptions.

D5·2 Pointwise Convergence Theorem

Let $f \in L^2_{\mathbb{C}}(\mathbb{R})$, and denote by f_m its projection on V_m. We have

$$f_m(t) = \int_{\mathbb{R}} q_m(x,t) f(x)\,\mathrm{d}x,$$

where q_m is the autoreproducing kernel of V_m, defined by (92). This kernel representation allows us to obtain pointwise convergence results, in the manner of Dirichlet's pointwise convergence analysis of Fourier series. We need some preliminary results on the kernel.

DEFINITION **D5.4.** *Let $\{\delta_m\}_{m\in\mathbb{Z}}$ be a sequence of functions $\delta_m : \mathbb{R}\times\mathbb{R} \to \mathbb{C}$. It is called a* quasi-positive delta sequence *if it satisfies the three following three conditions:*

(a) There exists a finite nonnegative constant K such that

$$\int_{\mathbb{R}} |\delta_m(x,t)|\mathrm{d}x < K, \textit{ for all } t \in \mathbb{R}, \textit{ all } m \in \mathbb{Z}. \tag{96}$$

(b) There exists a finite nonnegative constant c such that

$$\lim_{m\uparrow\infty} \int_{t-c}^{t+c} \delta_m(x,t)\,\mathrm{d}x = 1, \tag{97}$$

uniformly with respect to t in compact sets.

(c) For all $\gamma > 0$,

$$\lim_{m\uparrow\infty} \sup_{|x-t|\geq\gamma} |\delta_m(x,t)| = 0. \tag{98}$$

EXERCISE **D5.3.** *Show that Féjer's kernel sequence*

$$\delta_m(x,t) = \frac{\sin^2\left(\frac{m+1}{2}(x-t)\right)}{2\pi(m+1)\sin^2\left(\frac{1}{2}(x-t)\right)} 1_{[-\pi,+\pi]}(x-t)$$

is a quasi-positive delta sequence. Show that Dirichlet's kernel sequence

$$\delta_m(x,t) = \frac{\sin\left((m+\frac{1}{2})(x-t)\right)}{2\pi\sin\left(\frac{1}{2}(x-t)\right)} 1_{[-\pi,+\pi]}(x-t)$$

is not *a quasi-positive delta sequence.*

THEOREM **D5.1.** *If the MRA is r-smooth, the sequence $\{q_m\}m \in \mathbb{Z}$ defined by* (92) *is a quasi-positive delta sequence.*

Proof: We first prove property (a) of Definition D5.4:

$$\begin{aligned}\int_{\mathbb{R}} |q_m(x,t)|\mathrm{d}x &= \int_{\mathbb{R}} 2^m |q(2^m x, 2^m t)|\,\mathrm{d}x \\ &= \int_{\mathbb{R}} |q(x, 2^m t)|\,\mathrm{d}x \\ &\le C_2 \int_{\mathbb{R}} \frac{1}{(1+|x-2^m t|)^2}\mathrm{d}x \\ &\le C_2 \int_{\mathbb{R}} \frac{1}{(1+|x|)^2}\mathrm{d}x = K < \infty,\end{aligned}$$

where we have used inequality (90).

We now prove property (b) of Definition D5.4: Let $c > 0$ be finite. We have

$$\begin{aligned}\int_{t-c}^{t+c} q_m(x,t)\mathrm{d}x &= \int_{2^m(t-c)}^{2^m(t+c)} q(x, 2^m t)\mathrm{d}x \\ &= 1 - \int_{-\infty}^{2^m t - 2^m c} - \int_{2^m t + 2^m c}^{+\infty} = 1 - I_1 - I_2,\end{aligned}$$

where we have used (92) and (95). But

$$|I_1| \le C_2 \int_{-\infty}^{2^m t - 2^m c} \frac{1}{(1+|x-2^m t|)^2}\mathrm{d}x = C_2 \int_{-\infty}^{-2^m c} \frac{1}{(1+|x|)^2}\mathrm{d}x,$$

and this quantity tends to zero as m tends to infinity. A similar conclusion holds for I_2. Therefore, property (b) of Definition D5.4 is satisfied.

For property (c) of Definition D5.4, it suffices to observe that

$$2^m |q(2^m x, 2^m t)| \le C_2 2^m \frac{1}{(1+2^m|x-t|)^2}$$

in view of (90). ■

We can now state the main result.

THEOREM **D5.2.** *If $f \in L^1_{\mathbb{C}}(\mathbb{R}) \cap L^2_{\mathbb{C}}(\mathbb{R})$ is continuous on (a,b), then the projection $f_m = P_{V_m} f$ converges* pointwise *to f, uniformly on compact subintervals $[\alpha, \beta] \in (a,b)$, as $m \to \infty$.*

Proof: This is an immediate consequence of Theorem D5.1 and of the regularization lemma below. ■

LEMMA **D5.1.** *Let $\{\delta_m\}_{m\in\mathbb{Z}}$ be a quasi-positive delta sequence. Let $f \in L^1_{\mathbb{C}}(\mathbb{R})$ be continuous on (a, b), and define for all $m \in \mathbb{Z}$ the function f_m by*

$$f_m(t) = \int_{\mathbb{R}} \delta_m(x, t) f(x)\,\mathrm{d}x.$$

Then

$$\lim_{m\to\infty} f_m(t) = f(t)$$

uniformly on any compact subinterval $[\alpha, \beta] \in (a, b)$.

Proof: For $\gamma > 0$, write

$$\begin{aligned} f_m(t) = \int_{\mathbb{R}} \delta_m(x, t) f(x)\,\mathrm{d}x &= \int_{t-\gamma}^{t+\gamma} + \int_{t+\gamma}^{+\infty} + \int_{-\infty}^{t-\gamma} \\ &= f(t)\int_{t-\gamma}^{t+\gamma} \delta_m(x, t)\,\mathrm{d}x \\ &\quad + \int_{t-\gamma}^{t+\gamma} \delta_m(x, t)\,(f(t) - f(x))\,\mathrm{d}x + \left(\int_{t+\gamma}^{+\infty} + \int_{-\infty}^{t-\gamma}\right) \\ &= A + B + (C). \end{aligned}$$

Let $[\alpha, \beta] \in (a, b)$, $t \in [\alpha, \beta]$. Let c be as in (b) of D5.4. Choose γ such that $0 < \gamma < c$, $\beta + \gamma < b$, $\alpha - \gamma > a$. For any $0 < \varepsilon < 1$, further restrict γ so that $|f(x) - f(t)| < \varepsilon$ whenever $t \in [\alpha, \beta]$ and $|x - t| < \gamma$ (in which case both t and x are in a compact subinterval contained in (a, b), and we can then invoke the uniform continuity of f in this closed interval). We then have

$$|B| \le \varepsilon \int_{t-\gamma}^{t+\gamma} |\delta_m(x, t)|\,\mathrm{d}x \le \varepsilon \int_{\mathbb{R}} |\delta_m(x, t)|\,\mathrm{d}x \le \varepsilon K$$

and

$$|C| \le \sup_{|x-t|\ge\gamma} |\delta_m(x, t)| \int_{\mathbb{R}} |f(x)|\,\mathrm{d}x \le \varepsilon \int_{\mathbb{R}} |f(x)|\,\mathrm{d}x$$

for large enough m. Also, for large enough m,

$$\left|1 - \int_{t-\gamma}^{t+\gamma} \delta_m(x, t)\,\mathrm{d}x\right| \le \varepsilon$$

(use property (b) of the definition of quasi-delta sequences, and the fact that

$$\lim_{m\uparrow\infty} \int_{t+\gamma}^{t+c} \delta_m(x, t)\,\mathrm{d}x = 0$$

uniformly with $t \in \mathbb{R}$, and the same for the limit of $\int_{t-c}^{t-\gamma}$). Therefore, $|f(t) - A| \le \varepsilon$.

Putting all this together, we have for large enough m

$$|f(t) - f_m(t)| \leq |f(t) - A| + |B| + |C| \leq M\varepsilon + \varepsilon \int |f| + K\varepsilon,$$

where $M = \sup_{t\in[\alpha,\beta]} |f(t)|$. ■

D5·3 Regularity Properties of Wavelet Bases

In the wavelet expansion

$$f = \sum_{j,n} \langle f, \psi_{j,n} \rangle \psi_{j,n} = \sum_{j,n} d_{j,n} \psi_{j,n}\,,$$

where

$$\psi_{j,n}(t) = 2^{j/2} \psi(2^j t - n),$$

it is highly desirable from a numerical point of view that the coefficients $d_{j,n}$ decay rapidly as $|j|$, $|m| \to \infty$, thus ensuring fast convergence of the wavelet expansion. This is not the case, however, even for smooth functions (say, $f \in C^\infty \cap L^2$) if no further conditions are imposed on the mother wavelet ψ. To understand this and see what type of conditions ψ should satisfy, let us examine the asymptotic behavior of

$$d_{j,0} = 2^{j/2} \int_{\mathbb{R}} f(x)\psi(2^j x)^* \,\mathrm{d}x$$

as $j \to \infty$. Let $2^j = N$ and set

$$d(N) = \sqrt{N} \int_{\mathbb{R}} f(x)\psi(Nx)^* \,\mathrm{d}x$$

$$= \sqrt{N}\,\alpha(N).$$

A Taylor expansion of f (assumed to be C^∞) with Lagrange residue gives

$$\alpha(N) = \sum_{p=0}^{K} f^{(p)}(0) \int_{\mathbb{R}} \frac{x^p}{p!} \psi(Nx)^* \,\mathrm{d}x + \int_{\mathbb{R}} R_K(x)\psi(Nx)^* \,\mathrm{d}x,$$

where

$$R_K(x) = \int_0^x \frac{(x-t)^K}{K!} f^{(K+1)}(t)\,\mathrm{d}t.$$

We assume that the scaling function has a Fourier transform at 0 equal to 1, which implies that the mother wavelet has a null Fourier transform at 0 or, equivalently, that it integrates to 0. Therefore,

$$\alpha(N) = \frac{f'(0)}{N^2}\frac{\mu_1}{1!} + \frac{f''(0)}{N^3}\frac{\mu_2}{2!} + \cdots + \frac{f^{(K)}(0)}{N^{K+1}}\frac{\mu_K}{K!} + r_N(K),$$

where the μ_k's are the *wavelet moments*:

$$\mu_k = \int_{\mathbb{R}} x^k \psi(x)^* \, dx,$$

and the rest is readily bounded above by

$$r_N(K) \leq \frac{c}{N^{K+2}},$$

for some finite nonnegative c. In particular, a wavelet with moments that are zero up to order K implies

$$d(N) \leq \sqrt{N} \frac{c}{N^{K+2}}.$$

We shall see how the smoothness of ψ relates to moment conditions.

THEOREM D5.3. *Let $\psi \in S_r$ and assume that $\{\psi_{jk}\}_{j,k \in \mathbb{Z}}$ is a Hilbert basis of $L^2_{\mathbb{C}}(\mathbb{R})$, where*

$$\psi_{jk} = 2^{-j/2} \psi(2^{-j}x - k).$$

Then

$$\int_{\mathbb{R}} x^k \psi(x) \, dx = 0, \quad 0 \leq k \leq r. \tag{99}$$

Let N be a dyadic integer (that is, $N = 2^{-j_0}k_0$) such that $\psi(N) \neq 0$ (the existence of N follows by the density of dyadic integers in $\mathbb{R}$ and by the fact that ψ is continuous and not identically zero).

Let $j > 1$ be sufficiently large for $2^j N$ to be an integer. By orthogonality

$$\begin{aligned} 0 &= 2^j \int_{\mathbb{R}} \psi(x)\psi(2^j x - 2^j N) \, dx \\ &= \int_{\mathbb{R}} \psi(2^{-j}y + N)\psi(y) \, dy. \end{aligned} \tag{*}$$

Passing to the limit $j \to \infty$ gives, we have by dominated convergence

$$\psi(N) \int_{\mathbb{R}} \psi(y) \, dy = 0.$$

Therefore, (99) is proved for $k = 0$.

Suppose that (99) is true for $k = 1, \ldots, n-1$, where $n \leq r$. We have the Taylor expansion

$$\psi(x) = \sum_{k=0}^{n} \psi^{(k)}(N) \frac{(x-N)^k}{k!} + r_n(x) \frac{(x-N)^n}{n!},$$

where $r_n(x)$ is uniformly bounded. Choose N such that $\psi^{(n)}(N) \neq 0$. Substituting in (*), we obtain

$$0 = \int_{\mathbb{R}} \left\{ \sum_{k=0}^{n} \psi^{(k)}(N) \frac{(2^{-j}y)^k}{k!} + r_n(2^{-j}y + N) \frac{(2^{-j}y)^n}{n!} \right\} \psi(y) \, dy$$

$$= \psi^{(n)}(N) \int_{\mathbb{R}} 2^{-jn} \frac{y^n}{n!} \psi(y)\,\mathrm{d}y + \int_{\mathbb{R}} r_n(2^{-j}y + N) \frac{2^{-jn} y^n}{n!} \psi(y)\,\mathrm{d}y.$$

By dominated convergence, the last integral goes to zero as $j \to \infty$, and therefore,

$$\psi^{(n)}(N) \int_{\mathbb{R}} y^n \psi(y)\,\mathrm{d}y = 0. \blacksquare$$

Here is an apparent paradox relative to the moment conditions, and especially to the condition

$$\int_{\mathbb{R}} \psi(x)\,\mathrm{d}x = 0,$$

which is always satisfied and implies that the projection $f_m = P_{V_m} f$ satisfies

$$\int_{\mathbb{R}} f_m(x)\,\mathrm{d}x = 0, \tag{100}$$

a surprising fact at first glance, since the function f that is analyzed is in general *not* such that

$$\int_{\mathbb{R}} f_m(x)\,\mathrm{d}x = 0. \tag{101}$$

There is actually no contradiction since one cannot pass in the limit $m \to \infty$ in (100) to obtain (101): Convergence of f_m to f is in L^2, and this does not imply that

$$\lim_{m\uparrow\infty} \int_{\mathbb{R}} f_m(x)\,\mathrm{d}x = \int_{\mathbb{R}} f_m(x)\,\mathrm{d}x.$$

In Mallat's algorithm one first computes the projection $P_0 f$ that is the approximation of f at the resolution level 0, and then the coarser resolution approximations $P_j f$, $j \leq -1$. As we have just seen, the moment conditions on ψ are useful for the first part of the algorithm. For the second part fast decay of the coefficients

$$h_n = \sqrt{2} \int_{\mathbb{R}} \varphi(x)\varphi(2x - n)^*\,\mathrm{d}x$$

is needed for rapid numerical convergence. An ideal situation is when only a finite number of h_n are nonzero, which is guaranteed if the scaling function has compact support. Note that if this is the case, then the compactness of the scaling function carries over to the mother wavelet, and this is why one usually talks of *compact wavelets* rather than compact scaling functions.

Let us mention at this point that if we start from a Riesz basis of V_0, as in the method explained in Section D4·2, the compactness of w (there defined) does not imply compactness of the scaling function. In the face of this negative statement one needs to be reassured about the transmission of exponential decay from w to φ. As a matter of fact the situation is not too bad, and a result in this direction is, for example, Proposition 5.4.1 in [D3]. We end this section by showing how the decay of the scaling function is transmitted to the coefficients h_n. Localization of the scaling function can be taken in many related senses. We mentioned previously one of them, namely $\varphi \in S_r$. Another definition of localization could be

$$\int_{\mathbb{R}} (1 + |x|)^m |\varphi(x)|^2 \, dx < \infty, \quad \text{for all } m \in \mathbb{N}. \tag{102}$$

It follows from (102) that for finite constants c_m,

$$\int_{|x| \geq A} |\varphi(x)|^2 \, dx \leq \frac{c_m}{A^m}, \tag{103}$$

for all $A > 0$, $m \in \mathbb{N}$. By Schwarz's inequality,

$$\begin{aligned} |h_n| &\leq \sqrt{2} \left| \int_{|x| \geq A} \varphi(x)\varphi(2x - n) \, dx \right| \\ &\leq \sqrt{2} \left(\int_{|x| \geq A} |\varphi(2x - n)|^2 \, dx \right)^{1/2} + \sqrt{2} \left(\int_{|x| \leq A} |\varphi(x)|^2 \, dx \right)^{1/2}, \end{aligned}$$

and therefore with a proper choice of A, say $a = n$, in view of the tail majorization (103), we obtain

$$|h_n| \leq \frac{D_m}{(1 + n)^m} \quad \text{for all } m \in \mathbb{N}, \tag{104}$$

where the D_m are finite. Thus, the Fourier coefficients of m_0 are rapidly decaying and this implies that $m_0 \in C^\infty$.

The topic of compact wavelets is an important one, but it is rather technical. The interested reader is refered to [D3] for the detailed theory.

References

[D1] Blatter, C. (1998). *Wavelets, a Primer*, A. K. Peters: Natick, MA.

[D2] Chui, C.K. (1992). *An Introduction to Wavelets*, Academic Press: New York.

[D3] Daubechies, I. (1992). *Ten Lectures on Wavelets, CBSM–NSF Regional Conf. Series in Applied Mathematics*, SIAM: Philadelphia, PA.

[D4] Hernandez, E. and Weiss, G. (1996). *A First Course on Wavelets*, CRC Press: Boca Raton, FL.

[D5] Kahane, J.-P. and Lemarié-Rieusset, P.G. (1998). *Séries de Fourier et Ondelettes*, Cassini: Paris.

[D6] Mallat, S. (1998). *A Wavelet Tour of Signal Processing*, Wiley: New York.

[D7] Meyer, Y. (1993). *Wavelets Algorithms and Applications*, SIAM: Philadelphia, PA.

[D8] Vetterli, M. and Kovačević, J. (1995). *Wavelets and Sub-Band Coding*, Prentice-Hall: Englewood Cliffs, NJ.

[D9] Walter, G. (1994). *Wavelets and Other Orthogonal Systems with Applications*, CRC Press: Boca Raton, Fl.

Appendix

The Lebesgue Integral

Introduction

Integration is almost as old as mathematics. It is at least as old as Greek mathematics,[8] since Eudoxus and Archimedes used the exhaustion method to compute the volume of various solids, in particular, the pyramid and the cone.[9]

The modern theory of integration is intimately linked to Fourier series. Indeed, Bernhard Riemann (1826–1866) developed his theory of integration as a tool for studying Fourier series, the theme of his memoir of habilitation to professorship at the University of Gottingen. Also, Henri Lebesgue (1875–1941), who conceived his theory of integration in the period from 1902 to 1906, stated in a 1903 article: "*I am going to apply the notion of integral to the study of the trigonometric expansion of functions that are not integrable in the sense of Riemann.*"

The Riemann integral has a few weak points, the two main ones being that

---

[8]Sir Thomas Heath, *A History of Greek Mathematics; Vol. I: From Thales to Euclid*, Clarendon Press, Oxford, 1921; Dover edition, 1981.

[9]Exhaustion is the procedure by which we compute, for instance, the volume of the cone of height h and circular base of radius R, as the limit of a heap of circular tiles:

$$\lim_{n\uparrow\infty}\sum_{k=1}^{n}\pi\left(\frac{k}{n}R\right)^2\frac{h}{n}.$$

(1) The class of nonnegative functions which are Riemann-integrable is not large enough. Indeed, some functions have an "obvious" integral, and Riemann's integration theory denies it, while Lebesgue's theory recognizes it (see Example 9), and its stability properties under the limit operation are too weak.

(2) The Riemann integral is defined with respect to the Lebesgue measure (the "volume" in $\mathbb{R}^n$), whereas the Lebesgue integral can be defined with respect to a general abstract measure, a probability for instance.

The last advantage is an excellent argument to convince a student to invest a little time in the study of the Lebesgue integral, because the return is considerable. Indeed, the Lebesgue integral of the function f with respect to the measure μ (see the meaning in the first chapter), modestly denoted by

$$\int_X f(x)\,\mu(\mathrm{d}x),$$

contains a variety of mathematical objects, for instance, the usual Lebesgue integral on the line,

$$\int_{\mathbb{R}} f(x)\,\mathrm{d}x,$$

and also the Lebesgue volume integral. An infinite sum

$$\sum_{n\in\mathbb{Z}} f(n)$$

can also be viewed (with profit) as a Lebesgue integral with respect to the counting measure on $\mathbb{Z}$. The Stieltjes–Lebesgue integral

$$\int_{\mathbb{R}} f(x)\,\mathrm{d}F(x)$$

with respect to a function F of bounded variation, the expectation of a random variable Z:

$$E[Z]$$

are also in the scope of Lebesgue's integral. For the student who is reluctant to give up the expertise dearly acquired in the Riemann integral, it suffices to say that any Riemann-integrable function is also Lebesgue-integrable and that both integrals then coincide.

Is Lebesgue's theory hard to grasp? Not at all, because most of the results are very natural, and in that respect, the Lebesgue integral is much easier to manipulate correctly than the Riemann integral. A tedious (but not difficult) part is the step-by-step construction of the Lebesgue integral. However, if one just gives a summary of the main steps without going into the details, this is usually not a cause of frustration for the student interested in applications. The really difficult part is the proof of existence of certain measures, but students usually do not mind admitting such results. For instance, there is an existence theorem for the Lebesgue measure

ℓ (the "length") on $\mathbb{R}$. It says: There exists a unique measure ℓ on $\mathbb{R}$ that gives to the intervals $[a, b]$ the measure $b-a$. Of course, in order to understand what all the fuss is about, and what kind of mathematical subtleties hide behind such a harmless statement, we shall have to be more precise about the meaning of "measure". But when this is done, one is very much ready to approve the statement although the proof is not immediate. Of course, in this appendix, the proofs of such "obvious" results are not given. In fact, the goals of this appendix are to provide a tool and to give a few tips as to how to use it safely. The reader who has no previous knowledge of integration theory will therefore be very much in the situation of the new recipient of a driving license who takes the road in spite of her inexperience. Experience is best acquired on the road, and the main text contains many opportunities for the student to check her reflexes and to apply the rules that are briefly explained in the appendix. The student wishing to purchase good insurance is directed to the main companies, a few of which are listed in the bibliography of this appendix.

Farewell and bon voyage!

Measurable Functions and Measures

In this section, the basic steps in the construction of Lebesgue's integral are described, and the elementary properties of the integral are stated.

We first recall the notation: $\mathbb{N}$, $\mathbb{Z}$, $\mathbb{Q}$, $\mathbb{R}$, $\mathbb{C}$ are the sets of, respectively, integers, relative integers, rationals, real numbers, complex numbers; $\overline{\mathbb{R}} = \mathbb{R} \cup \{+\infty, -\infty\}$; $\mathbb{N}_+$ and $\mathbb{R}_+$ are the sets of positive integers and nonnegative real numbers; $\overline{\mathbb{R}}_+ = \mathbb{R}_+ \cup \{+\infty\}$.

Sigma-Fields

$\mathcal{P}(X)$ is the collection of all subsets of an arbitrary set X; card(X) is its cardinal, that is, the "number" of elements in it.

DEFINITION 1. *A family $\mathcal{X} \subset \mathcal{P}(X)$ of subsets of X is called a* sigma-field *on X if*

(α) $X \in \mathcal{X}$,
(β) $(A \in \mathcal{X}) \implies (\overline{A} \in \mathcal{X})$,
(γ) $(A_n \in \mathcal{X}$ for all $n \in \mathbb{N}) \implies (\cup_{n=0}^{\infty} A_n \in \mathcal{X})$.

One then says that $(X, \mathcal{X})$ is a measurable space.

Two extremal examples of sigma-fields on X are the *gross* sigma-field $\mathcal{X} = \{\varnothing, X\}$ and the *trivial* sigma-field $\mathcal{X} = \mathcal{P}(X)$.

The following situation is often encountered in measure theory: One has a collection of elementary sets, easy to describe mathematically, and one needs to define a sigma-field that contains these elementary sets and that is not too big.

DEFINITION 2. *The sigma-field generated by a collection of subsets $\mathcal{C} \subset \mathcal{P}(X)$ is, by definition, the smallest sigma-field on X containing all the sets in $\mathcal{C}$. It is denoted by $\sigma(\mathcal{C})$.*

Let $\{\mathcal{X}_i\}$, $i \in I$, be the collection of all sigma-fields on X containing $\mathcal{C}$. This collection is not empty, because $\mathcal{P}(X)$ belongs to it. Furthermore, one readily checks that $\cap_{i \in I} \mathcal{X}_i$ (by definition, the collection of subsets of X that belong to *all* the $\mathcal{X}_i$, $i \in I$) is a sigma-field. It contains $\mathcal{C}$ and, obviously, it is the smallest sigma-field containing $\mathcal{C}$. This proves the existence of $\sigma(\mathcal{C})$.

For the next definition, the reader not familiar with abstract topology may take $X = \mathbb{R}^n$ with the Euclidean topology.

DEFINITION 3. *Let X be a topological space and let $\mathcal{O}$ be the collection of open sets defining the topology. The sigma-field $\mathcal{B}(X) = \sigma(\mathcal{O})$ is called the* Borel sigma-field *on X associated with the given topology. A set $B \in \mathcal{B}(X)$ is called a* Borel set *of X.*

If $X = \mathbb{R}^n$ is endowed with the Euclidean topology, the Borel sigma-field $\mathcal{B}(\mathbb{R}^n)$ is denoted $\mathcal{B}^n$. For $n = 1$, we write $\mathcal{B}(\mathbb{R}) = \mathcal{B}$. For $I = \prod_{j=-1}^{n} I_j$, where I_j is a general interval of $\mathbb{R}$ (I is then called a *general rectangle* of $\mathbb{R}^n$), the Borel sigma-field $\mathcal{B}(I)$ on I consists of all the Borel sets contained in I.

THEOREM 1. *$\mathcal{B}(\mathbb{R}^n)$ is generated by the collection $\mathcal{C}$ of all rectangles of the type $\prod_{i=1}^{n}(-\infty, a_i]$, where $a_i \in \mathbb{Q}$ for all $i \in \{1, \ldots, n\}$.*

Measurable Functions

One of the central notions of Lebesgue's integration theory is that of a measurable function.

DEFINITION 4. *Let $(X, \mathcal{X})$ and $(E, \mathcal{E})$ be two measurable spaces. A function $f : X \mapsto E$ is said to be* measurable *with respect to $\mathcal{X}$ and $\mathcal{E}$ if*

$$f^{-1}(C) \in \mathcal{X} \quad \textit{for all } C \in \mathcal{E}.$$

This situation is denoted in various ways:

$$f : (X, \mathcal{X}) \mapsto (E, \mathcal{E}), \quad \text{or} \quad f \in \mathcal{E}/\mathcal{X}, \quad \text{or} \quad f \in \mathcal{X},$$

where the third notation will be used only when $(E, \mathcal{E}) = (I, \mathcal{B}(I))$, I being a general rectangle of $\mathbb{R}^n$, provided the context is clear enough as to the choice of I.

If $f : (X, \mathcal{X}) \mapsto (\mathbb{R}^k, \mathcal{B}^k)$ one says that f is a *Borel function* from X to $\mathbb{R}^k$. (However, this is not quite standard terminology; in the standard terminology, $(X, \mathcal{X})$ must be some $(\mathbb{R}^n, \mathcal{B}^n)$.)

Let $\overline{\mathcal{B}}$ be the sigma-field on $\overline{\mathbb{R}}$ generated by the intervals of type $(-\infty, a]$, $a \in \mathbb{R}$. A function $f : (X, \mathcal{X}) \mapsto (\overline{\mathbb{R}}, \overline{\mathcal{B}})$, where $(X, \mathcal{X})$ is an arbitrary measurable space, is called an *extended Borel function*, or simply a *Borel function*. As for functions $f : (X, \mathcal{X}) \mapsto (\mathbb{R}, \mathcal{B})$, they are called *real Borel functions*. In general, in a sentence such as "f is a Borel function defined on X," the sigma-field $\mathcal{X}$ is assumed to be the obvious one in the given context.

It seems difficult to prove measurability since most sigma-fields are not defined explicitly (see the definition of $\mathcal{B}^n$, for instance). However, the following result often simplifies the task.

THEOREM **2.** *Let $(X, \mathcal{X})$ and $(E, \mathcal{E})$ be two measurable spaces, where $\mathcal{E} = \sigma(\mathcal{C})$ for some collection $\mathcal{C}$ of subsets of E. Then $f : (X, \mathcal{X}) \mapsto (E, \mathcal{E})$ if and only if*

$$f^{-1}(C) \in \mathcal{X} \quad \text{for all } C \in \mathcal{C}.$$

One immediate application of this result is

EXAMPLE **1.** *Let $(X, \mathcal{X})$ be a measurable space and let $n \geq 1$ be an integer. Then $f = (f_1, \dots, f_n) : (X, \mathcal{X}) \mapsto (\mathbb{R}^n, \mathcal{B}^n)$ if and only if for all $a = (a_1, \dots, a_n) \in \mathbb{Q}^n$, $\{f \leq a\} \in \mathcal{X}$. (Here*

$$\{f \leq a\} := \{x \in \mathbb{R}^n : f_i(x) \leq a_i \text{ for all } 1 \leq i \leq n\}.)$$

The proof follows immediately from 2 and the definition of $\mathcal{B}^n$.

EXAMPLE **2.** *Let X and E be two topological spaces with respective Borel sigma-fields $\mathcal{B}(X)$ and $\mathcal{B}(E)$. Any continuous function $f : X \mapsto E$ is measurable with respect to $\mathcal{B}(X)$ and $\mathcal{B}(E)$.*

The above result is a direct consequence of Theorem 2 and of the abstract definition of continuity: $f : X \mapsto E$ is said to be continuous if $f^{-1}(O)$ is an open set of X whenever O is an open set of E.

Measurability is stable by composition:

THEOREM **3.** *Let $(X, \mathcal{X})$, $(Y, \mathcal{Y})$, and $(E, \mathcal{E})$ be three measurable spaces, and let $\varphi : (X, \mathcal{X}) \mapsto (Y, \mathcal{Y})$, $g : (Y, \mathcal{Y}) \mapsto (E, \mathcal{E})$. Then $f := g \circ \varphi : (X, \mathcal{X}) \mapsto (E, \mathcal{E})$.*

(This follows immediately from the definition of measurability.)

The next result shows that the set of Borel functions is stable by the "usual" operations.

THEOREM **4.** *(i) Let $f, g : (X, \mathcal{X}) \mapsto (\mathbb{R}, \mathcal{B})$. Then fg, $f + g$, $(f/g)1_{g \neq 0}$ are real Borel functions.*

(ii) Let $f_n : (X, \mathcal{X}) \mapsto (\mathbb{R}, \mathcal{B})$, $n \in \mathbb{N}$. Then $\liminf_{n\uparrow\infty} f_n$ *and* $\limsup_{n\uparrow\infty} f_n$ *are (possibly extended) Borel functions, and the set*

$$\{\limsup_{n\uparrow\infty} f_n = \liminf_{n\uparrow\infty} f_n\} = \{\exists \lim_{n\uparrow\infty} f_n\}$$

belongs to $\mathcal{X}$. In particular, if $\{\exists \lim_{n\uparrow\infty} f_n\} = X$, the function $\lim_{n\uparrow\infty} f_n$ is a (possibly extended) Borel function.

Measures

DEFINITION **5.** *Let $(X, \mathcal{X})$ be a measurable space and let $\mu : \mathcal{X} \mapsto [0, \infty]$ be a set function such that for any* denumerable *family $\{A_n\}_{n\geq 1}$ of* mutually disjoint *sets in $\mathcal{X}$,*

$$\mu\left(\sum_{n=1}^{\infty} A_n\right) = \sum_{n=1}^{\infty} \mu(A_n). \tag{105}$$

The set function μ *is called a* measure *on* $(X, \mathcal{X})$, *and* $(X, \mathcal{X}, \mu)$ *is called a* measure space.

Property (105) is the *sigma-additivity* property.

The following three properties are easy to check:

- $\mu(\emptyset) = 0$;
- $(A \subseteq B \text{ and } A, B \in \mathcal{X}) \implies (\mu(A) \leq \mu(B))$;
- $(A_n \in \mathcal{X} \text{ for all } n \in \mathbb{N}) \implies (\mu(\cup_{n=0}^{\infty} A_n) \leq \sum_{n=0}^{\infty} \mu(A_n))$.

EXAMPLE 3. *Let* $a \in X$. *The measure* ε_a *defined by* $\varepsilon_a(C) = 1_C(a)$ *is the* Dirac measure *at* $a \in X$. *The set function* $\mu : \mathcal{X} \mapsto [0, \infty]$ *defined by*

$$\mu(C) = \sum_{i=0}^{\infty} \alpha_i 1_{a_i}(C),$$

where $\alpha_i \in \overline{\mathbb{R}}_+$ *for all* $i \in \mathbb{N}$, *is a measure denoted* $\mu = \sum_{i=0}^{\infty} \alpha_i \varepsilon_{a_i}$.

EXAMPLE 4. *Let* $\{\alpha_n\}_{n \geq 1}$ *be a sequence of nonnegative numbers. The set function* $\mu : \mathcal{P}(\mathbb{Z}) \mapsto [0, \infty]$ *defined by* $\mu(C) = \sum_{n \in C} \alpha_n$ *is a measure on* $(\mathbb{Z}, \mathcal{P}(\mathbb{Z}))$. *If* $\alpha_n \equiv 1$ *we have the* counting measure ν *on* $\mathbb{Z}$, *where* $\nu(C) = \text{card}\,(C)$.

Next theorem introduces Lebesgue's measure.

THEOREM 5. *There exists one and only one measure* ℓ *on* $(\mathbb{R}, \mathcal{B})$ *such that*

$$\ell((a, b]) = b - a. \tag{106}$$

This measure is called the Lebesgue measure *on* $\mathbb{R}$.

DEFINITION 6. *Let* μ *be a measure on* $(X, \mathcal{X})$.
If $\mu(X) < \infty$ *the measure* μ *is called a* finite *measure.*
If $\mu(X) = 1$ *the measure* μ *is called a* probability *measure.*
If there exists a sequence $\{K_n\}_{n \geq 1}$ *of* $\mathcal{X}$ *such that* $\mu(K_n) < \infty$ *for all* $n \geq 1$, *and* $\cup_{n=1}^{\infty} K_n = X$, *the measure* μ *is called a* sigma-finite *measure.*
A measure μ *on* $(\mathbb{R}^n, \mathcal{B}^n)$ *such that* $\mu(C) < \infty$ *for all bounded Borel sets* C *is called a* Radon *measure.*

EXAMPLE 5. *The Dirac measure* ε_a *is a probability measure. The counting measure* ν *on* $\mathbb{Z}$ *is a sigma-finite measure. Any Radon measure on* $(\mathbb{R}^n, \mathcal{B}^n)$ *is sigma-finite. The Lebesgue measure is a Radon measure.*

Cumulative Distribution Function

DEFINITION 7. *A function* $F : \mathbb{R} \mapsto \mathbb{R}$ *is called a* cumulative distribution function *(c.d.f.) if the following properties are satisfied:*

1. F *is nondecreasing;*
2. F *is right-continuous;*
3. F *admits a left-hand limit, denoted* $F(x-)$, *at all* $x \in \mathbb{R}$.

EXAMPLE 6. *Let μ be a Radon measure on $(\mathbb{R}, \mathcal{B})$, and define*

$$F_\mu(t) = \begin{cases} \mu((0,t]) & \text{if } t \geq 0, \\ -\mu((t,0]) & \text{if } t < 0. \end{cases} \tag{107}$$

This is a c.d.f. (use next lemma), and, moreover,

$$F_\mu(b) - F_\mu(a) = \mu((a,b]),$$

$$F_\mu(a) - F_\mu(a-) = \mu(\{a\}).$$

F_μ is called the c.d.f. of μ.

From the last formula, we deduce that any point set $\{a\}, a \in \mathbb{R}$ has null Lebesgue measure, and therefore, any countable subset of $\mathbb{R}$ ($\mathbb{Q}$, for instance), has null Lebesgue measure.

The following lemma features the sequential continuity properties of measures.

LEMMA 1. *Let $(X, \mathcal{X}, \mu)$ be a measure space. Let $\{A_n\}_{n\geq 1}$ be a non-decreasing (that is, $A_n \subseteq A_{n+1}$ for all $n \geq 1$) sequence of $\mathcal{X}$. Then*

$$\mu\left(\bigcup_{n=1}^{\infty} A_n\right) = \lim_{n\uparrow\infty} \uparrow \mu(A_n). \tag{108}$$

Let $\{B_n\}_{n\geq 1}$ be a nonincreasing (that is, $B_{n+1} \subseteq B_n$ for all $n \geq 1$) sequence of $\mathcal{X}$ such that $\mu(B_{n_0}) < \infty$ for some $n_0 \in \mathbb{N}_+$. Then

$$\mu\left(\bigcap_{n=1}^{\infty} B_n\right) = \lim_{n\downarrow\infty} \downarrow \mu(B_n). \tag{109}$$

Proof: We shall prove (108). This equality follows directly from sigma-additivity since

$$\mu(A_n) = \mu(A_1) + \sum_{i=1}^{n-1} \mu(A_{i+1} - A_i)$$

and

$$\mu\left(\bigcup_{n=1}^{\infty} A_n\right) = \mu(A_1) + \sum_{i=1}^{\infty} \mu(A_{i+1} - A_i).$$

The proof of (109) is left to the reader. ■

The necessity of the condition $\mu(B_{n_0}) < \infty$ for some n_0 is illustrated by the following counterexample. Let ν be the counting measure on $\mathbb{Z}$, and for all $n \geq 1$ define $B_n = \{i \in \mathbb{Z} : |i| \geq n\}$. Then $\nu(B_n) = +\infty$ for all $n \geq 1$, and

$$\nu\left(\bigcap_{n=1}^{\infty} B_n\right) = \nu(\varnothing) = 0.$$

We now state a fundamental existence result, which generalizes Theorem 5.

THEOREM **6.** *Let $F : \mathbb{R} \mapsto \mathbb{R}$ be a c.d.f.. There exists a unique measure μ on $(\mathbb{R}, \mathcal{B})$ such that $F_\mu = F$.*

This result is easily stated, but it is not trivial. It is typical of the existence results, which answer the following type of question: Let $\mathcal{C}$ be a collection of subsets of X with $\mathcal{C} \subset \mathcal{X}$, where $\mathcal{X}$ is the sigma-field on X generated by $\mathcal{C}$. Given a set function $u : \mathcal{C} \mapsto [0, \infty]$, does there exist a measure μ on $(X, \mathcal{X})$ such that $\mu(C) = u(C)$ for all $C \in \mathcal{C}$, and is it unique?

The reason why such results are nontrivial is that the sigma-field generated by $\mathcal{C}$ is not explicitly constructed. It is therefore not easy to say what $\mu(C)$ should be when one does not really know what a typical $C \in \mathcal{X}$ should look like!

Negligible Sets

A very important concept in measure and integration theory is that of negligible sets, with the correlated notion of "almost everywhere."

DEFINITION **8.** *A μ-negligible set is a set contained in a set $N \in \mathcal{X}$ such that* $\mu(N) = 0$.

*Let $\mathcal{P}$ be some property relative to the elements $x \in X$, where $(X, \mathcal{X}, \mu)$ is a measure space. One says that $\mathcal{P}$ holds μ-*almost everywhere (μ-a.e.) *if the set* $\{x \in X : x$ does not satisfy $\mathcal{P}\}$ *is a μ-negligible set.*

For instance, if f and g are two Borel functions defined on X, the expression

$$f \leq g \quad \mu\text{-a.e.}$$

means that

$$\mu(\{x : f(x) > g(x)\}) = 0.$$

The following result is easy to prove.

THEOREM **7.** *A countable union of μ-negligible sets is a μ-negligible set.*

The following result is used several times in the main text.

THEOREM **8.** *If two continuous functions $f, g : \mathbb{R} \mapsto \mathbb{R}$ are ℓ-a.e. equal, they are everywhere equal.*

Proof: Let $t \in \mathbb{R}$ be such that $f(t) \neq g(t)$. For any $c > 0$, there exists $s \in [t - c, t + c]$ such that $f(s) = g(s)$. (Otherwise, the set $\{t; f(t) \neq g(t)\}$ would contain the whole interval $[t - c, t + c]$ and therefore could not be of null Lebesgue measure.) Therefore, one can construct a sequence $\{t_n\}_{n \geq 1}$ converging to t and such that $f(t_n) = g(t_n)$ for all $n \geq 1$. Letting n tend to ∞ yields $f(t) = g(t)$, a contradiction. ■

The Integral

Having defined measures and measurable functions, we are ready to construct the abstract Lebesgue integral.

The Simple Case

A Borel function $f : (X, \mathcal{X}) \mapsto (\mathbb{R}, \mathcal{B})$ of the type

$$f(x) = \sum_{i=1}^{k} a_i \, 1_{A_i}(x),$$

where $k \in \mathbb{N}_+$, $a_1, \ldots, a_k \in \mathbb{R}$, $A_1, \ldots, A_k \in \mathcal{X}$, is called an *elementary Borel function* (defined on X).

The following result is the key to the construction of the Lebesgue integral:

THEOREM 9. *Let $f : (X, \mathcal{X}) \mapsto (\overline{\mathbb{R}}, \overline{\mathcal{B}})$ be a nonnegative Borel function. There exists a nondecreasing sequence $\{f_n\}_{n\geq 1}$ of nonnegative elementary Borel functions that converges pointwise to f.*

Proof: Take

$$f_n(x) = \sum_{k=0}^{n2^{-n}-1} k2^{-n} \, 1_{A_{k,n}}(x),$$

where

$$A_{k,n} = \{x \in X : k2^{-n} < f(x) \leq (k+1)2^{-n}\}.$$

■

For any nonnegative elementary Borel function $f : (X, \mathcal{X}) \mapsto (\mathbb{R}, \mathcal{B})$ of the form

$$f(x) = \sum_{i=1}^{k} a_i 1_{A_i}(x),$$

where $a_i \in \mathbb{R}_+$, $A_i \in \mathcal{X}$ for all $i \in \{1, \ldots, k\}$, one defines the integral of f with respect to μ, denoted

$$\int_X f \, \mathrm{d}\mu, \quad \text{or} \quad \int_X f(x) \, \mu(\mathrm{d}x), \quad \text{or} \quad \mu(f),$$

by

$$\int_X f \, \mathrm{d}\mu = \sum_{i=1}^{k} a_i \, \mu(A_i). \tag{110}$$

The Case of Nonnegative Measurable Functions

If $f : (X, \mathcal{X}) \mapsto (\overline{\mathbb{R}}, \overline{\mathcal{B}})$ is nonnegative, the integral is defined by

$$\int_X f \, \mathrm{d}\mu = \lim_{n\uparrow\infty} \uparrow \int_X f_n \, \mathrm{d}\mu, \tag{111}$$

where $\{f_n\}_{n\geq 1}$ is a nondecreasing sequence of nonnegative elementary Borel functions $f_n : (X, \mathcal{X}) \mapsto (\mathbb{R}, \mathcal{B})$ such that $\lim_{n\uparrow\infty} \uparrow f_n = f$. This definition can be shown to be consistent, in that the integral so defined is independent of the choice of the approximating sequence. Note that the quantity (111) is nonnegative and can be infinite. It can be shown that if $f \leq g$, where $f, g : (X, \mathcal{X}) \mapsto (\overline{\mathbb{R}}, \overline{\mathcal{B}})$ are nonnegative, then

$$\int_X f \, d\mu \leq \int_X g \, d\mu.$$

In particular, if

$$f^+ = \max(f, 0) \quad \text{and} \quad f^- = \max(-f, 0),$$

we have

$$f = f^+ - f^-, \qquad f^\pm \leq |f|,$$

and therefore,

$$\int_X f^\pm \, d\mu \leq \int_X |f| \, d\mu. \tag{112}$$

Integrable Functions

DEFINITION 9. *A measurable function* $f : (X, \mathcal{X}) \mapsto (\overline{\mathbb{R}}, \overline{\mathcal{B}})$ *is called a* μ-integrable function *if*

$$\int_X |f| \, d\mu < \infty. \tag{113}$$

In this case (see (112)) the right-hand side of

$$\int_X f \, d\mu := \int_X f^+ \, d\mu - \int_X f^- \, d\mu \tag{114}$$

is meaningful and defines the left-hand side. Moreover, the integral of f with respect to μ defined in this way is finite.

The integral can be defined for nonintegrable functions in certain circumstances; for example, it is defined in the nonnegative case even when the function is not integrable. More generally, if $f : (X, \mathcal{X}) \mapsto (\overline{\mathbb{R}}, \overline{\mathcal{B}})$ is such that at least one of the integrals $\int_X f^+ \, d\mu$ or $\int_X f^- \, d\mu$ is finite, one defines

$$\int_X f \, d\mu = \int_X f^+ \, d\mu - \int_X f^- \, d\mu. \tag{115}$$

This leads to one of the forms "finite minus finite," "finite minus infinite," and "infinite minus finite." The case $\mu(f^+) = \mu(f^-) = +\infty$ is rigorously excluded from the definition, because it leads to the indeterminate form "infinite minus infinite."

Counting Measure and Dirac Measure

The results in the following two examples are easy to prove.

EXAMPLE 7. *Any function* $f : \mathbb{Z} \mapsto \mathbb{R}$ *is measurable with respect to* $\mathcal{P}(\mathbb{Z})$ *and* $\mathcal{B}$. *With the measure* μ *defined in Example 4, and with* $f \geq 0$ *for instance, we have*

$$\mu(f) = \sum_{n=1}^{\infty} \alpha_n f(n).$$

EXAMPLE 8. *Let* ε_a *be the Dirac measure at point* $a \in X$. *Then any* $f : (X, \mathcal{X}) \mapsto (\mathbb{R}, \mathcal{B})$ *is* ε_a*-integrable, and*

$$\varepsilon_a(f) = f(a).$$

Elementary Properties of the Integral

First, recall that for all $A \in \mathcal{X}$,

$$\int_X 1_A \, \mathrm{d}\mu = \mu(A). \tag{116}$$

Also, recall the notation $\int_A f \, \mathrm{d}\mu$ for $\int_X 1_A f \, \mathrm{d}\mu$.

THEOREM 10. *Let* $f, g : (X, \mathcal{X}) \mapsto (\overline{\mathbb{R}}, \overline{\mathcal{B}})$ *be* μ*-integrable functions, and let* $a, b \in \mathbb{R}$. *Then*

(a) $af + bg$ *is* μ*-integrable and* $\mu(af + bg) = a\mu(f) + b\mu(g)$,
(b) if $f = 0$ μ*-a.e., then* $\mu(f) = 0$; *If* $f = g$ μ*-a.e., then* $\mu(f) = \mu(g)$,
(c) if $f \leq g$ μ*-a.e., then* $\mu(f) \leq \mu(g)$,
(d) $|\mu(f)| \leq \mu(|f|)$,
(e) if $f \geq 0$ μ*-a.e. and* $\mu(f) = 0$, *then* $f = 0$ μ*-a.e.*,
(f) if $\mu(1_A f) = 0$ *for all* $A \in \mathcal{X}$, *then* $f = 0$ μ*-a.e.*.
(g) if f *is* μ*-integrable, then* $|f| < \infty$ μ*-a.e.*

For a complex Borel function $f : X \mapsto \mathbb{C}$ (i.e., $f = f_1 + i f_2$, where $f_1, f_2 : (X, \mathcal{X}) \mapsto (\mathbb{R}, \mathcal{B})$) such that $\mu(|f|) < \infty$, one defines

$$\int_X f \, \mathrm{d}\mu = \int_X f_1 \, \mathrm{d}\mu + i \int_X f_2 \, \mathrm{d}\mu. \tag{117}$$

The extension to complex Borel functions of the properties (a), (b), (d), and (f) in Theorem 10 is immediate.

Riemann and Lebesgue

The following result tells us that all the time spent learning about the Riemann integral has not been in vain.

THEOREM 11. *Let* $f : (\mathbb{R}, \mathcal{B}) \mapsto (\mathbb{R}, \mathcal{B})$ *be Riemann-integrable. Then it is Lebesgue-integrable with respect to* ℓ, *and the Lebesgue integral is equal to the Riemann integral.*

EXAMPLE 9. *The converse is not true: The function* f *defined by* $f(x) = 1$ *if* $x \in \mathbb{Q}$ *and* $f(x) = 0$ *if* $x \notin \mathbb{Q}$ *is a Borel function, and it is Lebesgue-integrable with its integral equal to zero because* $\{f \neq 0\} = \mathbb{Q}$, *has* ℓ*-measure zero. However,* f *is not Riemann-integrable.*

EXAMPLE **10.** *The function* $f : (\mathbb{R}, \mathcal{B}) \mapsto (\mathbb{R}, \mathcal{B})$ *defined by*

$$f(x) = \frac{x}{1+x^2}$$

does not have a Lebesgue integral, because

$$f^+(x) = \frac{x}{1+x^2} 1_{[0,\infty)}(x) \quad \text{and} \quad f^-(x) = -\frac{x}{1+x^2} 1_{(-\infty,0]}(x)$$

have infinite Lebesgue integrals. However, it has a generalized Riemann integral

$$\lim_{A \uparrow \infty} \int_{-A}^{+A} \frac{x}{1+x^2} \, dx = 0.$$

Limits Under the Integral

The three main results that we need to know in this book are the Lebesgue theorems (when can we interchange the order of limit and integration?), the Tonnelli–Fubini theorems (when can we interchange the order of integration in a multiple integral?), and the theorem of completeness of L^2.

Lebesgue, Fatou, and Beppo Levi

The following result of Beppo Levi is often called the *monotone convergence* theorem.

THEOREM **12.** *Let* $f_n : (X, \mathcal{X}) \mapsto (\overline{\mathbb{R}}, \overline{\mathcal{B}})$, $n \geq 1$, *be such that*

(i) $f_n \geq 0$ μ*-a.e.,*

(ii) $f_{n+1} \geq f_n$ μ*-a.e.*

Then there exists a nonnegative function $f : (X, \mathcal{X}) \mapsto (\overline{\mathbb{R}}, \overline{\mathcal{B}})$ *such that*

$$\lim_{n \uparrow \infty} \uparrow f_n = f \quad \mu\text{-a.e.}$$

and

$$\int_X f \, d\mu = \lim_{n \uparrow \infty} \uparrow \int_X f_n \, d\mu.$$

The next result is a useful technical tool called *Fatou's lemma.*

THEOREM **13.** *Let* $f_n : (X, \mathcal{X}) \mapsto (\overline{\mathbb{R}}, \overline{\mathcal{B}})$, $n \geq 1$, *be such that* $f_n \geq 0$ μ*-a.e. for all* $n \geq 1$. *Then*

$$\int_X (\liminf_{n \uparrow \infty} f_n) \, d\mu \leq \liminf_{n \uparrow \infty} \left(\int_x f_n \, d\mu \right). \tag{118}$$

The *dominated convergence theorem* is also called the Lebesgue theorem:

THEOREM **14.** *Let* $f_n : (X, \mathcal{X}) \mapsto (\overline{\mathbb{R}}, \overline{\mathcal{B}})$, $n \geq 1$, *be such that, for some function* $f : (X, \mathcal{X}) \mapsto (\overline{\mathbb{R}}, \overline{\mathcal{B}})$ *and some* μ*-integrable function* $g : (X, \mathcal{X}) \mapsto (\overline{\mathbb{R}}, \overline{\mathcal{B}})$,

(i) $\lim_{n \uparrow \infty} f_n = f$, μ*-a.e.,*

(ii) $|f_n| \le |g|$ *μ-a.e. for all* $n \ge 1$.

Then

$$\int_X f\,\mathrm{d}\mu = \lim_{n\uparrow\infty}\left(\int_X f_n\,\mathrm{d}\mu\right).$$

The results in Theorems 12 and 14 ensure that under certain circumstances limit and integration may be interchanged (that is, $\mu(\lim f_n) = \lim \mu(f_n)$). The classical counterexample that shows this is not always true is the following:

EXAMPLE **11.** *For* $(X, \mathcal{X}, \mu) = (\mathbb{R}, \mathcal{B}, \ell)$, *define*

$$f_n(x) = 0 \quad \text{if} \quad |x| > \frac{1}{n},$$

$$f_n(x) = n^2 x + n \quad \text{if} \quad -\frac{1}{n} \le x \le 0,$$

$$f_n(x) = -n^2 x + n \quad \text{if} \quad 0 \le x \le \frac{1}{n}.$$

One has

$$\lim_{n\uparrow\infty} f_n(x) = 0 \quad \text{if} \quad x \neq 0,$$

that is, $\lim_{n\uparrow\infty} f_n = 0$ *μ-a.e. Therefore,* $\mu(\lim_{n\uparrow\infty} f_n) = 0$. *However,* $\mu(f_n) = 1$ *for all* $n \ge 1$.

Differentiation Under the Integral

A very useful application of the dominated convergence theorem is the theorem of differentiation under the integral sign. Let $(X, \mathcal{X}, \mu)$ be a measure space and let $(a, b) \subset \mathbb{R}$. Let $f : (a, b) \times X \mapsto \mathbb{R}$ and, for all $t \in (a, b)$, define $f_t : X \mapsto \mathbb{R}$ by $f_t(x) = f(t, x)$. Assume that for all $t \in (a, b)$, f_t is measurable with respect to $\mathcal{X}$, and define, if possible, the function $I : (a, b) \mapsto \mathbb{R}$ by the formula

$$I(t) = \int_X f(t, x)\,\mu(\mathrm{d}x). \tag{119}$$

THEOREM **15.** *Assume that for μ-almost all x the function $t \rightsquigarrow f(t, x)$ is continuous at $t_0 \in (a, b)$ and that there exists a μ-integrable function $g : (X, \mathcal{X}) \mapsto (\overline{\mathbb{R}}, \overline{\mathcal{B}})$ such that $|f(t, x)| \le |g(x)|$ μ-a.e. for all t in a neighborhood V of t_0. Then $I : V \mapsto \mathbb{R}$ is well defined and is continuous at t_0. Furthermore, assume that*

(α) $t \to f(t, x)$ is continuously differentiable on V for μ-almost all x,
(β) For some μ-integrable function $h : (X, \mathcal{X}) \mapsto (\overline{\mathbb{R}}, \overline{\mathcal{B}})$

$$\left|\frac{\partial f}{\partial t}(t, x)\right| \le |h(x)| \quad \mu\text{-a.e.}.$$

Then I is differentiable at t_0 and

$$I'(t_0) = \int_X \frac{\partial f}{\partial t}(t_0, x)\,\mu(\mathrm{d}x). \tag{120}$$

Proof: Let $\{t_n\}_{n\geq 1}$ be a sequence in $V \setminus \{t_0\}$ such that $\lim_{n\uparrow\infty} t_n = t_0$, and define $f_n(x) = f(t_n, x)$, $f(x) = f(t_0, x)$. Then, by dominated convergence,

$$\lim_{n\uparrow\infty} I(t_n) = I(t_0).$$

Also,

$$\frac{I(t_n) - I(t_0)}{t_n - t_0} = \int_X \frac{f(t_n, x) - f(t_0, x)}{t_n - t_0} \mu(\mathrm{d}x),$$

and for some $\theta \in (0, 1)$, possibly depending upon n,

$$\left| \frac{f(t_n, x) - f(t_0, x)}{t_n - t_0} \right| \leq \left| \frac{\partial f}{\partial t}(t_0 + \theta(t_n - t_0), x) \right|.$$

The latter quantity is bounded by $|h(x)|$. Therefore, by dominated convergence,

$$\lim_{n\uparrow\infty} \frac{I(t_n) - I(t_0)}{t_n - t_0} = \int_X \left(\lim_{n\uparrow\infty} \frac{f(t_n, x) - f(t_0)}{t_n - t_0} \right) \mu(\mathrm{d}x)$$

$$= \int_X \frac{\partial f}{\partial t}(t_0, x)\, \mu(\mathrm{d}x).$$ ■

The Fubini Theorem

Product Measures

Let $(X_1, \mathcal{X}_1, \mu_1)$ and $(X_2, \mathcal{X}_2, \mu_2)$ be two measure spaces where μ_1 and μ_2 are sigma-finite measures.

Define the product set $X = X_1 \times X_2$ and the *product sigma-field* $\mathcal{X} = \mathcal{X}_1 \times \mathcal{X}_2$, where by definition the latter is the smallest sigma-field on X containing all sets of the form $A_1 \times A_2$, where $A_1 \in \mathcal{X}_1$, $A_2 \in \mathcal{X}_2$.

THEOREM **16.** *There exists unique measure* μ *on* $(X_1 \times X_2, \mathcal{X}_1 \times \mathcal{X}_2)$ *such that*

$$\mu(A_1 \times A_2) = \mu_1(A_1)\mu_2(A_2) \tag{121}$$

for all $A_1 \in \mathcal{X}_1$, $A_2 \in \mathcal{X}_2$.

The measure μ is the *product measure of* μ_1 *and* μ_2, and is denoted $\mu_1 \otimes \mu_2$.

The above result extends in an obvious manner to a finite number of sigma-finite measures.

EXAMPLE **12.** *The typical example of a product measure is the Lebesgue measure on the space* $(\mathbb{R}^n, \mathcal{B}^n)$: *It is the unique measure* ℓ^n *on that space that is such that*

$$\ell^n\left(\prod_{i=1}^n A_i\right) = \prod_{i=1}^n \ell(A_i) \quad \text{for all } A_1, \ldots, A_n \in \mathcal{B}.$$

Tonnelli and Fubini

Going back to the situation with two measure spaces (the case of a finite number of measure spaces is similar) we have the following result:

THEOREM **17.** *Let* $(X_1, \mathcal{X}_1, \mu_1)$ *and* $(X_1, \mathcal{X}_2, \mu_2)$ *be two measure spaces in which* μ_1 *and* μ_2 *are* sigma-finite. *Let* $(X, \mathcal{X}, \mu) = (X_1 \times X_2, \mathcal{X}_1 \times \mathcal{X}_2, \mu_1 \otimes \mu_2)$.
(A) Tonelli. If f *is nonnegative, then, for* μ_1*-almost all* x_1*, the function* $x_2 \to f(x_1, x_2)$ *is measurable with respect to* $\mathcal{X}_2$*, and*

$$x_1 \to \int_{X_2} f(x_1, x_2)\, \mu_2(dx_2)$$

is a measurable function with respect to $\mathcal{X}_1$*. Furthermore,*

$$\int_X f\, d\mu = \int_{X_1} \left[\int_{X_2} f(x_1, x_2)\, \mu_2(dx_2) \right] \mu_1(dx_1). \tag{123}$$

(B) Fubini. If f *is* μ*-integrable, then, for* μ_1*-almost all* x_1*, the function* $x_2 \to f(x_1, x_2)$ *is* μ_2*-integrable and* $x_1 \to \int_{X_2} f(x_1, x_2)\, \mu_2(dx_2)$ *is* μ_2*-integrable, and* (123) *is true.*

In this text we shall refer to the global result as the *Fubini–Tonelli* theorem.

Part (A) says that one can integrate a nonnegative Borel function in any order of its variables. Part (B) says that the same is true of an arbitrary Borel function if that function is μ-integrable. In general, in order to apply Part (B), one must use Part (A) with $f = |f|$ to ascertain whether or not $\int |f|\, d\mu < \infty$.

EXAMPLE **13.** *Consider the function* f *defined on* $X_1 \times X_2 = (1, \infty) \times (0, 1)$ *by the formula*

$$f(x_1, x_2) = e^{-x_1 x_2} - 2e^{-2x_1 x_2}.$$

We have

$$\int_{(1,\infty)} f(x_1, x_2)\, dx_1 = \frac{e^{-x_2} - e^{-2x_2}}{x_2}$$

$$= h(x_2) \geq 0,$$

$$\int_{(0,1)} f(x_1, x_2)\, dx_2 = -\frac{e^{-x_1} - e^{-2x_1}}{x_1}$$

$$= -h(x_1).$$

However,

$$\int_0^1 h(x_2)\, dx_2 \neq \int_1^\infty (-h(x_1))\, dx_1,$$

since $h \geq 0$ ℓ*-a.e. on* $(0, \infty)$*. We therefore see that successive integrations yield different results according to the order in which they are performed. As a matter of fact,* $f(x_1, x_2)$ *is* not *integrable on* $(0, 1) \times (1, \infty)$.

Integration by Parts

THEOREM **18.** *Let μ_1 and μ_2 be two sigma-finite measures on $(\mathbb{R}, \mathcal{B})$. For any interval $(a, b) \subset \mathbb{R}$,*

$$\mu_1((a,b])\mu_2((a,b]) = \int_{(a,b]} \mu_1((a,t])\,\mu_2(\mathrm{d}t) + \int_{(a,b]} \mu_2((a,t))\,\mu_1(\mathrm{d}t). \quad (124)$$

Observe that in the first integral we have $(a, t]$ (closed on the right), whereas in the second integral we have (a, t) (open on the right).

Proof: The proof consists of computing the μ-measure of the square $(a, b] \times (a, b]$ in two ways. The first one is obvious and gives the left-hand side of (124). The second one consists of observing that $\mu((a, b] \times (a, b]) = \mu(D_1) + \mu(D_2)$, where $D_1 = \{(x, y); a < y \le b, a < x \le y\}$ and $D_2 = (a, b] \times (a, b] \setminus D_1$. Then $\mu(D_1)$ and $\mu(D_2)$ are computed using Tonelli's theorem. For instance,

$$\mu(D_1) = \int_{\mathbb{R}} \left(\int_{\mathbb{R}} 1_{D_1}(x, y)\mu_1(\mathrm{d}x) \right) \mu_2(\mathrm{d}y)$$

and

$$\int_{\mathbb{R}} 1_{D_1}(x, y)\mu_1(\mathrm{d}x) = \int_{\mathbb{R}} 1_{\{a<x\le y\}}\mu_1(\mathrm{d}x) = \mu_1((a, y]).$$ ■

Let μ be a Radon measure on $(\mathbb{R}, \mathcal{B})$ and let F_μ be its c.d.f. The notation

$$\int_{\mathbb{R}} g(x)\, F_\mu(\mathrm{d}x)$$

stands for $\int_{\mathbb{R}} g(x)\,\mu(\mathrm{d}x)$. When this integral is used, it is usually called the *Lebesgue–Stieltjes integral* of g with respect to F_μ. With this notation, (124) becomes

$$F_1(b)F_2(b) - F_1(a)F_1(b) = \int_{(a,b]} F_1(x)\,\mathrm{d}F_2(x) + \int_{(a,b]} F_2(x-)\,\mathrm{d}F_1(x), \quad (125)$$

where $F_i := F_{\mu_i}$ $(i = 1, 2)$. This is the Lebesgue–Stieltjes version of the integration by parts formula of calculus.

The Spaces L^p

For a given $p \ge 1$, $L^p_{\mathbb{C}}(\mu)$ is, roughly speaking (see the details below), the collection of complex-valued Borel functions f defined on X such that $\int_X |f|^p \,\mathrm{d}\mu < \infty$. We shall see that it is a complete normed vector space over $\mathbb{C}$, that is, a Banach space. Of special interest to Fourier analysis is the case $p = 2$, since $L^2_{\mathbb{C}}(\mu)$ has additional structure that makes of it a Hilbert space.

Let $(X, \mathcal{X}, \mu)$ be a measure space and let f, g be two complex-valued Borel functions defined on X. The relation $\mathcal{R}$ defined by

$$(f\mathcal{R}g) \Longleftrightarrow (f = g \;\; \mu\text{-a.e.})$$

is an equivalence relation, and we shall denote the equivalence class of f by $\{f\}$.

Note that for any $p > 0$ (using property b of 10),

$$(f\mathcal{R}g) \implies \left(\int_X |f|^p \, d\mu = \int_X |g|^p \, d\mu\right).$$

The operations $\times$, $+$, $*$, and multiplication by a scalar $\alpha \in \mathbb{C}$ are defined on the equivalence class by

$$\{f\} + \{g\} = \{f + g\}, \quad \{f\}^* = \{f^*\}, \quad \alpha\{f\} = \{\alpha f\}, \quad \{f\}\{g\} = \{fg\}.$$

The first equality means that $\{f\} + \{g\}$ is, by definition, the equivalence class consisting of the functions $f + g$, where f and g are abritrary members of $\{f\}$ and $\{g\}$, respectively. A similar interpretation holds for the other equalities.

By definition, for a given $p \geq 1$, $L^p_{\mathbb{C}}(\mu)$ is the collection of equivalence classes $\{f\}$ such that $\int_X |f|^p \, d\mu < \infty$. Clearly, it is a vector space over $\mathbb{C}$ (for the proof recall that

$$\left(\frac{|f| + |g|}{2}\right)^p \leq \frac{1}{2}|f|^p + \frac{1}{2}|g|^p$$

since $t \to t^p$ is a convex function when $p \geq 1$).

In order to avoid cumbersome notation, in this section and in general whenever we consider L^p-spaces, we shall write f for $\{f\}$. This abuse of notation is harmless since two members of the same equivalence class have the same integral if that integral is defined. Therefore, using loose notation,

$$L^p_{\mathbb{C}}(\mu) = \{f : \int_X |f|^p \, d\mu < \infty\}. \tag{126}$$

The following is a simple and often used observation.

THEOREM **19.** *Let p and q be positive real numbers such that $p > q$. If the measure μ on $(X, \mathcal{X}, \mu)$ is finite, then $L^p_{\mathbb{C}}(\mu) \subseteq L^q_{\mathbb{C}}(\mu)$. In particular, $L^2_{\mathbb{C}}(\mu) \subseteq L^1_{\mathbb{C}}(\mu)$.*

Proof: From the inequality $|a|^q \leq 1 + |a|^p$, true for all $a \in \mathbb{C}$, it follows that $\mu(|f|^q) \leq \mu(1) + \mu(|f|^p)$. Since $\mu(1) = \mu(\mathbb{R}) < \infty$, $\mu(|f|^q) < \infty$ whenever $\mu(|f|^p) < \infty$. ■

Hölder's and Minkowski's Inequalities

THEOREM **20.** *Let p and q be positive real numbers different from 1 such that*

$$\frac{1}{p} + \frac{1}{q} = 1$$

(p and q are then said to be conjugate), and let $f, g : (X, \mathcal{X}) \mapsto (\overline{\mathbb{R}}, \overline{\mathcal{B}})$ be nonnegative. Then, we have Hölder's inequality

$$\int_X fg \, d\mu \leq \left[\int_X f^p \, d\mu\right]^{1/p} \left[\int_X g^q \, d\mu\right]^{1/q}. \tag{127}$$

In particular, if $f, g \in L^2_{\mathbb{C}}(\mathbb{R})$, then $fg \in L^1_{\mathbb{C}}(\mathbb{R})$.

Proof: Let

$$A = \left(\int_X (f^p)\,\mathrm{d}\mu\right)^{1/p},$$

$$B = \left(\int_X (g^q)\,\mathrm{d}\mu\right)^{1/q}.$$

We may assume that $0 < A < \infty$, $0 < B < \infty$, because otherwise Hölder's inequality is trivially satisfied.

Define $F = f/A$, $G = g/A$, so that

$$\int_X F^p\,\mathrm{d}\mu = \int_X G^q\,\mathrm{d}\mu = 1.$$

The inequality

$$F(x)G(x) \le \frac{1}{p}F(x)^p + \frac{1}{q}G(x)^q \tag{*}$$

is trivially satisfied if x is such that $F(x) = 0$ or $G(x) = 0$. If $F(x) > 0$ and $G(x) > 0$, define

$$s(x) = p\ln(F(x)), \qquad t(x) = q\ln(G(x)).$$

From the convexity of the exponential function and the assumption that $1/p + 1/q = 1$,

$$e^{s(x)/p+t(x)/q} \le \frac{1}{p}e^{s(x)} + \frac{1}{q}e^{t(x)},$$

and this is precisely the inequality (*). Integrating this inequality yields

$$\int_X (FG)\,\mathrm{d}\mu \le \frac{1}{p} + \frac{1}{q} = 1,$$

and this is just (127).

THEOREM **21.** *Let $p \ge 1$ and let $f, g : (X, \mathcal{X}) \mapsto (\overline{\mathbb{R}}, \overline{\mathcal{B}})$ be nonnegative and such that*

$$\int_X f^p\,\mathrm{d}\mu < \infty, \qquad \int_X g^p\,\mathrm{d}\mu < \infty.$$

Then, we have Minkowski's inequality

$$\left[\int_X (f+g)^p\right]^{1/p} \le \left[\int_X f^p\,\mathrm{d}\mu\right]^{1/p} + \left[\int_X g^p\,\mathrm{d}\mu\right]^{1/p}. \tag{128}$$

Proof: For $p = 1$ the inequality is obvious. Therefore, assume $p > 1$. From Hölder's inequality,

$$\int_X f(f+g)^{p-1}\,\mathrm{d}\mu \le \left[\int_X f^p\,\mathrm{d}\mu\right]^{1/p}\left[\int_X (f+g)^{(p-1)q}\right]^{1/q}$$

and

$$\int_X g(f+g)^{p-1}\,d\mu \le \left[\int_X g^p\,d\mu\right]^{1/p}\left[\int_X (f+g)^{(p-1)q}\right]^{1/q}.$$

Adding together the above two inequalities and observing that $(p-1)q = p$, we obtain

$$\int_X (f+g)^p\,d\mu \le \left[\int_X (f+g)^p\right]^{1/q}\left(\left[\int_X f^p\,d\mu\right]^{1/p} + \left[\int_X g^p\,d\mu\right]^{1/p}\right).$$

One may assume that the right-hand side of (128) is finite and that the left-hand side is positive (otherwise the inequality is trivial). Therefore, $\int_X (f+g)^p\,d\mu \in (0,\infty)$. We may therefore divide both sides of the last display by $\left[\int_X (f+g)^p\,d\mu\right]^{1/q}$. Observing that $1 - 1/q = 1/p$ yields the desired inequality (128).

For the last assertion of the theorem, take $p = q = 2$. ■

THEOREM **22.** *Let $p \ge 1$. The mapping $\nu_p : L^p_{\mathbb{C}}(\mu) \mapsto [0,\infty)$ defined by*

$$\nu_p(f) = \left(\int_X |f|^p\,d\mu\right)^{1/p} \tag{129}$$

defines a norm on $L^p_{\mathbb{C}}(\mu)$.

Proof: Clearly, $\nu_p(\alpha f) = |\alpha|\nu_p(f)$ for all $\alpha \in \mathbb{C}$, $f \in L^p_{\mathbb{C}}(\mu)$.

Also, $(\nu_p(f) = 0) \Longleftrightarrow (\int_X |f|^p\,d\mu)^{1/p} = 0 \implies (f = 0)$.
Finally, $\nu_p(f+g) \le \nu_p(f) + \nu_p(g)$ for all $f, g \in L^p_{\mathbb{C}}(\mu)$, by Minkowski's inequality.

Therefore, ν_p is a norm. ■

Riesz–Fischer Theorem

We shall denote $\nu_p(f)$ by $\|f\|_p$. Thus $L^p_{\mathbb{C}}(\mu)$ is a normed vector space over $\mathbb{C}$, with the norm $\|\cdot\|_p$ and the induced distance

$$d_p(f,g) = \|f-g\|_p.$$

THEOREM **23.** *Let $p \ge 1$. The distance d_p makes of $L^p_{\mathbb{C}}$ a* complete *normed space.*

In other words, $L^p_{\mathbb{C}}(\mu)$ is a *Banach space* for the norm $\|\cdot\|_p$.

Proof: To show completeness one must prove that for any Cauchy sequence $\{f_n\}_{n\ge1}$ of $L^p_{\mathbb{C}}(\mu)$ there exists $f \in L^p_{\mathbb{C}}(\mu)$ such that $\lim_{n\uparrow\infty} d_p(f_n, f) = 0$.

Since $\{f_n\}_{n\ge1}$ is a Cauchy sequence (that is, $\lim_{m,n\uparrow\infty} d_p(f_n, f_m) = 0$), one can select a subsequence $\{f_{n_i}\}_{i\ge1}$ such that

$$d_p(f_{n_{i+1}} - f_{n_i}) \le 2^{-i}. \tag{*}$$

Let

$$g_k = \sum_{i=1}^{k} \left|f_{n_{i+1}} - f_{n_i}\right|,$$

$$g = \sum_{i=1}^{\infty} |f_{n_{i+1}} - f_{n_i}|.$$

By (*) and Minkowski's inequality we have $\|g_k\|_p \le 1$. Fatou's lemma applied to the sequences $\{g_k^p\}_{k\ge 1}$ gives $\|g\|_p \le 1$. In particular, any member of the equivalence class of g is finite μ-almost everywhere, and therefore

$$f_{n_1}(x) + \sum_{i=1}^{\infty} (f_{n_{i+1}}(x) - f_{n_i}(x))$$

converges absolutely for μ-almost all x. Call this limit $f(x)$ (set $f(x) = 0$ when this limit does not exist). Since

$$f_{n_1} + \sum_{i=1}^{k-1} (f_{n_{i+1}} - f_{n_i}) = f_{n_k},$$

we see that

$$f = \lim_{k\uparrow\infty} f_{n_k} \quad \mu\text{-a.e.}$$

One must show that f is the limit in $L^p_{\mathbb{C}}(\mu)$ of $\{f_{n_k}\}_{k\ge 1}$. Let $\varepsilon > 0$. There exists an integer $n = N(\varepsilon)$ such that $\|f_n - f_m\|_p \le \varepsilon$ whenever $m, n \ge N$. For all $m > N$, by Fatou's lemma we have

$$\int_X |f - f_m|^p \, d\mu \le \liminf_{i\to\infty} \int_X |f_{n_i} - f_m|^p \, d\mu \le \varepsilon^p.$$

Therefore, $f - f_m \in L^p_{\mathbb{C}}(\mu)$, and consequently, $f \in L^p_{\mathbb{C}}(\mu)$. It also follows from the last inequality that

$$\lim_{m\to\infty} \|f - f_m\|_p = 0.$$

■

Terminology. For $p \ge 1$, $L^p_{\mathbb{C}}(\mu)$ is a Banach space (a complete normed vector space) over $\mathbb{C}$. This phrase will implicitly assume that the norm is defined as in (129). When μ is the Lebesgue measure on $\mathbb{R}^n$, we write $L^p_{\mathbb{C}}(\mathbb{R}^n)$ instead of $L^p_{\mathbb{C}}(\mu)$ (with a slight symbolic inconsistency).

In the proof of Theorem 23 we obtained the following result.

THEOREM **24.** *Let $\{f_n\}_{n\ge 1}$ be a convergent sequence in $L^p_{\mathbb{C}}(\mu)$, where $p \ge 1$, and let f be the limit.*

A subsequence $\{f_{n_i}\}_{i\ge 1}$ can then be chosen such that

$$\lim_{i\uparrow\infty} f_{n_i} = f \quad \mu\text{-a.e.} \tag{130}$$

Note that the statement in (130) is about functions and not about equivalence classes. The functions thereof are *any* members of the corresponding equivalence class. In particular, since when a given sequence of functions converges μ-a.e. to two functions, these two functions are necessarily equal μ-a.e.

THEOREM **25.** *If $\{f_n\}_{n\geq 1}$ converges both to f in $L^p_{\mathbb{C}}(\mu)$ and to g μ-a.e., then $f = g$ μ-a.e.*

A most interesting special case of L^p-space, particularly in view of its relevance to signal processing, is when $p = 2$. In this case

THEOREM **26.** *$L^2_{\mathbb{C}}(\mu)$ is a complete normed space with respect to the norm*

$$\|f\| = \left[\int_X |f|^2 \, \mathrm{d}\mu\right]^{1/2}.$$

This norm is derived from a Hermitian product, namely,

$$\langle f, g\rangle = \int_X fg^* \, \mathrm{d}\mu$$

in the sense that

$$\|f\|^2 = \langle f, f\rangle.$$

$L^2_{\mathbb{C}}(\mu)$ is a Hilbert space over $\mathbb{C}$ (see Section C1·1).

Approximation Theorems

We now quote the approximation results used in the main text.

THEOREM **27.** *Let $f \in L^p_{\mathbb{C}}(\mathbb{R})$, $p \geq 1$. There exists a sequence $\{f_n\}_{n\geq 1}$ of continuous functions $f_n : \mathbb{R} \mapsto \mathbb{C}$ with compact support that converges to f in $L^p_{\mathbb{C}}(\mathbb{R})$.*

(To have compact support means, for a continuous function, to be null outside some closed bounded interval.)

THEOREM **28.** *Let $f \in L^p_{\mathbb{C}}(\mathbb{R})$, $p \geq 1$. There exists a sequence $\{f_n\}_{n\geq 1}$ of functions $f_n : \mathbb{R} \mapsto \mathbb{C}$ which are finite linear combinations of indicator functions of intervals, that converges to f in $L^p_{\mathbb{C}}(\mathbb{R})$.*

THEOREM **29.** *Let $f \in L^1_{\mathbb{C}}([-\pi, +\pi])$ be a 2π-periodic function (that is, $f(t) = f(t+2\pi)$ for all $t \in \mathbb{R}$, and $\int_{-\pi}^{+\pi} |f(t)| \, \mathrm{d}t < \infty$). There exists a sequence $\{f_n\}_{n\geq 1}$ of functions $f_n : \mathbb{R} \mapsto \mathbb{C}$ with continuous derivatives that converges to f in $L^1_{\mathbb{C}}([-\pi, +\pi])$.*

References

[D1] de Barra, G. (1981). *Measure Theory and Integration*, Ellis Horwood: Chichester.

[D2] Halmos, P.R. (1950). *Measure Theory*, Van Nostrand: New York.

[D3] Royden, H.L. (1988). *Real Analysis*, 3rd ed., MacMillan: London.

[D4] Rudin, W. (1966). *Real and Complex Analysis*, McGraw-Hill: New York.

[D5] Taylor, A.E. (1965). *General Theory of Functions and Integration*, Blaisdell, Waltham, MA, Dover edition, 1985.

Glossary of Symbols

$\mathcal{P}(X)$, the collection of all subsets of set X.

card (X), or $|X|$, the cardinal of set X; the number of elements in X.

$\mathbb{N}$, the integers.

$\mathbb{N}_+$, the positive integers.

$\mathbb{Z}$, the relative integers.

$\mathbb{R}$, the reals.

$\mathbb{R}_+$, the positive reals.

$\mathbb{C}$, the complex numbers.

z^*, the complex conjugate of $z \in \mathbb{C}$.

$(a, b]$, interval of $\mathbb{R}$ open to the left, closed to the right; and similar notation for the other types of intervals.

$\mathrm{Re}(z)$, the real part of $z \in \mathbb{C}$.

$\mathrm{Im}(z)$, the imaginary part of $z \in \mathbb{C}$.

$f : \mathbb{R} \mapsto \mathbb{C}$, a function from $\mathbb{R}$ to $\mathbb{C}$; equivalent notation: f, $f(\cdot)$, $f(t)$.

$f^{(n)}$, the nth derivative of f; $f^{(0)} = f$.

$f * g$, the convolution product of f and g:

$$(f * g)(t) = \int_{\mathbb{R}} f(t-s)\, g(s)\, ds = \int_{\mathbb{R}} g(t-s)\, f(s)\, ds.$$

f^{*n}, the nth convolution product of f by itself:

$$f^{*0} = f;\, f^{*(n+1)} = f * f^{*n}.$$

1_A, the indicator function of a set A; $1_A(t) = 1$ if $t \in A$, $= 0$ otherwise.

$f_T(t) = f(t)1_{[0,T]}(t)$.

$\text{sinc}\,(t) = \sin(\pi t)/\pi t$, the cardinal sine function.

$rect_T\,(t) = 1_{[-\frac{T}{2},+\frac{T}{2}]}(t)$, the rectangle function.

ℓ, the Lebesgue measure on $\mathbb{R}$; $\ell([a,b]) = b - a$.

a.e., almost everywhere with respect to the Lebesgue measure.

μ-a.e., almost everywhere with respect to the measure μ.

$L^p_{\mathbb{C}}(\mathbb{R})$, the set (equivalence classes) of measurable functions $f : \mathbb{R} \mapsto \mathbb{C}$ such that $\int_{\mathbb{R}} |f(t)|^p\, dt < \infty$.

$L^p_{\mathbb{C}}([a,b])$, the set (equivalence classes) of measurable functions $f : [a,b] \mapsto \mathbb{C}$ such that $\int_a^b |f(t)|^p\, dt < \infty$.

$L^p_{\mathbb{C},loc}(\mathbb{R})$, the set (equivalence classes) of measurable functions $f : \mathbb{R} \mapsto \mathbb{C}$ such that $f(t)1_{[a,b]}(t) \in L^p_{\mathbb{C}}(\mathbb{R})$ for all $[a,b] \subset \mathbb{R}$.

L^p_{loc}, short for $L^p_{\mathbb{C},loc}(\mathbb{R})$.

$\ell^2_{\mathbb{R}}(\mathbb{Z})$, the set of complex sequences $\{x_n\}_{n\in\mathbb{Z}}$ such that $\sum_{n\in\mathbb{Z}} |x_n|^2 < \infty$.

C^n, the set of $n$times continuously differentiable functions $f : \mathbb{R} \mapsto \mathbb{C}$.

C^∞, the set of infinitely differentiable functions $f : \mathbb{R} \mapsto \mathbb{C}$.

C^0, the set of continuous functions $f : \mathbb{R} \mapsto \mathbb{C}$.

C^0_c, the set of continuous functions $f : \mathbb{R} \mapsto \mathbb{C}$ with bounded support.

$C([0,T])$, the set of continuous functions $f : [0,T] \mapsto \mathbb{C}$.

$\mathcal{D}$, the set of test functions $\varphi : \mathbb{R} \mapsto \mathbb{C}$; C^∞ and compact support.

$\mathcal{D}'$, the set of distributions on $\mathbb{R}$; the set of linear forms on $\mathcal{D}$.

$\mathcal{S}$, the set of functions $f : [0,T] \mapsto \mathbb{C}$ in C^∞ and with all its derivatives rapidly decreasing.

$\mathcal{S}_r$, the set of functions $f : [0,T] \mapsto \mathbb{C}$ in C^r and with all its derivatives up to order r rapidly decreasing.

$\mathcal{S}'$, the set of tempered distributions on $\mathbb{R}$; the set of linear forms on $\mathcal{S}$.

$\langle x, y\rangle_H$, the Hermitian product of $x, y \in H$, H Hilbert space.

$\|x\|_H = \langle x, x\rangle_H)^{1/2}$, the norm of $x \in H$, H Hilbert space.

$x \perp y$; x is orthogonal to y; $\langle x, y\rangle_H = 0$.

$G^\perp$, the orthogonal complement of G.

P_G, the orthogonal projection on G.

$c_n(f) = (1/2\pi)\int_0^{2\pi} f(t)e^{-int}\,dt$, the nth Fourier coefficient of f.

$S_n(f) = \sum_{-n}^{+n} c_k(f)e^{+ikt}$, the Fourier series.

$S(f) = \sum_{n\in\mathbb{Z}} c_n(f)e^{+int}$, the formal Fourier series development.

$\hat{f}(\nu) = \int_{\mathbb{R}} f(t)e^{-2i\pi\nu t}\,dt$, the Fourier transform of f.

$H(z) = \sum_{n\in\mathbb{Z}} h_n z^n$, the z-transform of $\{h_n\}_{n\in\mathbb{Z}}$.

$\varphi_{j,n}(t) = 2^{j/2}\varphi(2^j t - n)$.

Index